Intellectual Property in the Food Technology Industry

T0184480

Ryan W. O'Donnell · John J. O'Malley ·
Randolph J. Huis · Gerald B. Halt, Jr.

Intellectual Property in the Food Technology Industry

Protecting Your Innovation

 Springer

Ryan W. O'Donnell
Volpe and Koenig, P.C.,
Philadelphia, PA,USA
rodonnell@volpe-koenig.com

John J. O'Malley
Volpe and Koenig, P.C.,
Philadelphia, PA,USA
jomalley@volpe-koenig.com

Randolph J. Huis
Volpe and Koenig, P.C.,
Philadelphia, PA,USA

Gerald B. Halt, Jr.
Volpe and Koenig, P.C.,
Philadelphia, PA,USA

The authors are all shareholders at the intellectual property law firm of Volpe and Koenig, P.C., and have extensive industry and legal experience. Volpe and Koenig provides guidance on matters relating to patents, trademarks, copyrights, trade secrets, e-commerce, technology joint ventures, non-disclosure agreements, technology acquisitions, licensing and litigation. In addition to food technology, the firm has experience in the electronics, consumer goods, wireless technology, mechanical, medical, chemical, biotechnical and pharmaceutical industries. For more information about the authors or Volpe and Koenig, P.C. please visit www.volpe-koenig.com.

ISBN 978-0-387-77388-9   ISBN 978-0-387-77389-6 (eBook)
DOI 10.1007/978-0-387-77389-6

Library of Congress Control Number: 2008926489

Printed on acid-free paper

springer.com

Acknowledgments

The authors would like to thank everyone at the law firm of Volpe and Koenig, P.C. for providing support and encouragement in preparing this book. The authors would also like to individually thank Anthony S. Volpe for his contribution to the intellectual property enforcement and litigation chapter, C. Frederick Koenig III for his contribution to the copyright chapter, and Stephen B. Schott for his contribution to the overall organization and editing of this book.

Contents

Part II Implementing IP Practices and Procedures

Introduction

Creativity can create economic value. This maxim holds true equally for the food industry as for other industries. Such value may come from a new innovation, edging out competitors in a market, creating a revenue stream where there was none, or increasing market reputation. This book provides an introduction to intellectual property law, as applied to the food technology industry. This area of law provides the legal framework for bridging creativity and the value that may come from it. Through the proper use of intellectual property law, one has a much better chance of transforming creativity into economic value.

Intellectual property law recognizes a creator's rights in ideas, innovations, and goodwill. Being intangible, intellectual property differs from real property (land) or personal property (your possessions) that are secured, controlled, and protected using physical means such as fences, locks, alarms, and guards. Because intellectual property is a product of the mind, there is often no easy way to build a "fence" around it. Consider one of the most valuable trademarks in the world: Coca-Cola®. The Coca-Cola Company could not protect this mark with a physical fence. It is intellectual property law that provides a legal fence of trademark protection to protect the goodwill of the Coca-Cola® trademark.

There are a variety of intellectual property pitfalls that await the unwary. Different rules apply to different types of intellectual property (IP). Accordingly, you may forfeit your rights if you do not take the appropriate measures to secure and protect them. Thus, it is important to understand the types of IP protection and the respective rules that govern each type of IP.

1. *Patent*: Patents may be granted for the invention of any new and useful process, machine, manufacture or composition of matter, or any new useful improvement thereof. A patent is a property right that grants the inventor or owner the right to exclude others from making, using, selling, or offering to sell the invention as defined by the patent's claims in the United States for a limited period of time.

2. *Trademark*: A trademark is a word, phrase, symbol, or design, or combination of words, phrases, symbols, or designs which identifies and distinguishes the source of the goods or services of one party from those of others. Trademarks promote competition by giving products corporate identity and marketing leverage.
3. *Copyright*: Copyrights protect original works of authorship fixed in a tangible medium of expression. Copyrighted works include literary, dramatic, and musical compositions, movies, pictures, paintings, sculptures, computer programs, etc. Copyright protects the expression of an idea, but not the idea itself.
4. *Trade Secret*: Generally, a trade secret is any formula, manufacturing process, method of business, technical know-how, etc. that gives its holder a competitive advantage and is not generally known. The legal definition of a trade secret and the protection afforded to a trade secret owner varies from state to state.

The table below highlights some of the attributes of and distinctions between these different types of IP:

	Patent	Trade Secret	Trademark	Copyright
Underlying theory	Limited monopoly to encourage innovation in exchange for disclosure of invention to the public	Protects proprietary and sensitive business information against improper acquisition	Used to identify the source of a product or service to consumers, and to distinguish the source of products or services from other sources	Limited monopoly to encourage the authorship of works
Subject matter	Processes, machines, articles of manufacture, compositions of matter, asexually reproduced plants, designs for articles of manufacture. Laws of nature, mathematical algorithms, natural phenomena, mental steps, etc. are not patentable	Formulas, patterns, compilations, programs, devices, methods, techniques, processes, etc. that derive independent economic value from being "secret"	Trademarks, service marks, trade names, certification marks, collective marks, trade dress	Literary, musical, choreographic, dramatic, and artistic works *limited by* idea/expression dichotomy (no protection for ideas, systems, methods, procedures); no protection for facts/research
Legal source	Patent Act (35 USC §100 *et seq.*)	State statutes (e.g., Uniform Trade Secrets Act); common law	Lanham Act (15 USC §§1051–1127); common law	Copyright Act (17 USC 101 *et seq.*); some limited common law
Legal standards	Must be patentable subject matter, novel, non-obvious, and useful	Information must not be generally known or readily available. Reasonable efforts to maintain secrecy must be taken	Must be distinctive or carry a secondary meaning (for descriptive and geographic marks), and must be used in commerce	Must be an original work of authorship fixed in a tangible medium

	Patent	Trade Secret	Trademark	Copyright
Scope of rights	Exclusive right to prevent others from making, using, selling or offering to sell the subject matter of the patent	Protection against improper acquisition by others	Exclusive right to use the mark in within a particular territory depending on the type of trademark protection	Exclusive right to perform, display, reproduce, or make derivative works
Term	20 years from application filing date	No time limitation. Protection is available as long as kept secret	No time limitation. Protection is available as long as used in commerce	Generally, the term is the life of the author plus 70 years. For works of corporate authorship, the term is 120 years after creation or 95 years after publication, whichever endpoint is earlier
Enforcement/remedies	File suit for patent infringement. Remedy can be damages (lost profits or reasonable royalty) and injunctive relief	File suit for misappropriation, conversion, or breach of contract. Remedy is typically damages	File suit for trademark infringement. Remedies can include injunctive relief, accounting for profits, destruction of goods, etc.	File suit for infringement. Remedies include injunctive relief, destruction of infringing goods, and damages (actual or profits or statutory damages)

Patents, trademarks, trade secrets, and copyrights all have a strong presence in the food technology industry. Trademarks are perhaps the most common means of IP protection in the food industry. Companies often invest millions in advertising and marketing their brands in order to build up goodwill and consumer loyalty towards their products. Many of the "supermarks", trademarks that have achieved a level of famousness to be considered a household name, come from food products and services. Marks such as Coca-Cola®, Cheerios®, and McDonalds® are instantly recognized by the general consuming public worldwide as a designation of source and associated with an expected level of quality. Many purchasing decisions in the food technology industry are based on brand name alone, and that is why so many food technology companies pursue trademark protection, as summarized below:

Class name	Current live applications and registered trademarks	Trademarks registered in 2006	Trademark applications filed in 2006
Meats and processed foods	44,924	3161	5404
Staple foods	68,400	5207	8596
Natural agricultural products	24,503	1753	2798
Light beverages	25,519	1826	4028
Wines and spirits	29,254	2801	4901

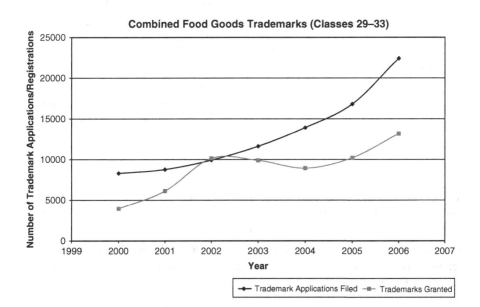

As shown, there is a steady increase in trademark filing in the food industry over the period from 1999 to 2006. Companies are increasingly investing in their brand and reputation by filing for trademark protection. It is equally important for a company to secure patent and trade secret rights in its research and development.

Once secured, a company can enforce its intellectual property rights against a competitor. Several notable examples are summarized below.

(1) In *McNeil-Ppc, Inc. v. Merisant Co.*, Civ. No. 04-1090, 2004 US Dist. LEXIS 27733 (D.P.R. 2004), McNeil, the manufacture of a Splenda®, filed an action against Merisant, the manufacturer of Equal® and Nutrasweet®, for trade dress infringement and false advertising. McNeil sought a preliminary injunction preventing Merisant from marketing a new no-calorie sweetener in packaging that was confusingly similar to that of Splenda®. The court granted McNeil's motion for a preliminary injunction, holding that the McNeil was highly likely to prevail on the merits of its trade dress claim under the Lanham Act. The court's order granting a preliminary injunction included a product recall, and required Merisant to post a bond.

(2) In *Kemin Foods, L.C. v. Pigmentos Vegetales Del Centro S.A. de C.V.*, 464 F.3d 1339 (Fed. Cir. 2006), Kemin Foods filed a patent infringement suit against Pigmentos for infringement of two patents directed to purified lutein that is extracted from plants. The defendant filed a counterclaim seeking a declaratory judgment that Kemin's patents were invalid and unenforceable. The court of appeals affirmed the district court's holding that the patent claims were not invalid, and that Kemin Foods was entitled to damages based on defendant's infringement of its patents.

(3) In *Michael Foods v. Papetti's Hygrade Egg Prods.*, 1994 US App. LEXIS 18323 (Fed. Cir. 1994), the plaintiff filed an action against the defendant, an egg company, for infringement of patents directed to egg product pasteurization. The patent claims a method for ultrapasteurizing a liquid egg product. The plaintiff was successful in enforcing its patent against the defendant on summary judgment, which was upheld by the court of appeals.

The above cases were filed in federal court to enforce federal IP rights. Another commonly used option is to file suit in the International Trade Commission (ITC) to prevent the importation of articles that infringe a valid and enforceable US patent, registered copyright, or trademark. Some food-related investigations brought in the ITC include patent infringement claims against foreign manufacturers for plastic food containers and plastic grocery bags. *See Plastic Food Containers*, ITC Inv. No. 337-TA-514 (2005), *Plastic*

Grocery and Retail Bags, ITC Inv. No. 337-TA-492 (2004). In both cases, the complainant was successful in obtaining an exclusion order to prevent importation of infringing articles into the United States. In another case, a soft drink company, Kola Columbiana, was able to obtain limited exclusion orders against various Columbian companies infringing Kola Columbiana's trademark and trade dress. *Soft Drinks and Their Containers*, ITC Inv. No. 337-TA-321 (1991). In yet another food-related case, Yamasa Enterprises, a California-based manufacturer of fish and seafood products, obtained a limited exclusion order to prevent several foreign companies from infringing its registered trademark. *Asian-Style Kamaboko Fish Cakes*, ITC Inv. No. 337-TA-378 (1996).

This book illustrates how intellectual property rights can apply by presenting the example of a fictitious company, Tastewell Industries, and its development of a new cheese product that consists of a mixture of certain processed cheese and various fruits.

Part I of this book provides a comprehensive overview of the most common forms of intellectual property rights. Part II provides guidelines for how food technology companies can properly secure, implement, leverage, and enforce their intellectual property rights.

Part I
Overview of Intellectual Property Rights

Chapter 1
Patents

1.1 What Is a Patent and Why Apply for a Patent?

Congress shall have power ... To promote the progress of science and useful arts, by securing for limited times to authors and inventors the exclusive right to their respective writings and discoveries.

United States Constitution, Article I, § 8.

A US patent is a contract between the United States and the inventor(s) in which the owner is granted a limited monopoly to exclude others from making, using, selling, offering for sale, or importing a patented invention into the United States for a period of time during the term of the patent. In exchange for these exclusive rights, the inventor is required to disclose the full and complete details of the invention to the public. The theory behind the patent system is that if the public has access to complete inventive disclosures, it will develop new and better ways of solving the same problems.

The patent monopoly has some limitations. A patent does not give an owner the right to make, use, or sell an invention. For example, a patent owner can be prevented from selling its patented invention if a competitor's earlier patent covers some part of the patented invention. Further, a US patent is not enforceable outside the United States; each country offers its own patent protections.

The patent right to exclude others from making, using, selling, offering for sale, or importing the patented invention creates barriers for competitors to enter the market. Such barriers often facilitate licensing arrangements where some of the patent rights can be separated. For example, a company can grant a license to one company to *make* a patented invention, while granting a second license to another company to *use or sell* the patented invention. Developing a strong portfolio of patent rights (i.e., barriers to entry) can be attractive to investors or can create new business opportunities by reducing the risks of competition.

R.W. O'Donnell et al., *Intellectual Property in the Food Technology Industry,*
DOI: 10.1007/978-0-387-77389-6_1, © Springer Science+Business Media, LLC 2008

Example: Tastewell Industries has a patent for its cheese and fruit product. One of the ingredients in the product is a sweetener that is covered by a patent and owned by Sweet Ingredients, Inc.

While Tastewell has the right to exclude others from making, using, selling, offering for sale, or importing mixed cheese and fruit products within the scope of Tastewell's patent rights, Tastewell will need a license from Sweet Ingredients if it intends to make, use, sell, or offer the product for sale. Tastewell would normally receive an implied license if it buys the sweetener from Sweet Ingredients. If Tastewell obtains the sweetener from another source, Tastewell will want assurances from its vender that the sweetener is covered by an appropriate license.

1.2 Types of Patents and Applications

There are several types of US patents issued by the United States Patent and Trademark Office (USPTO): utility, design, and plant patent. Utility patents are the most common and protect functional innovations including "any new and useful process, machine, manufacture, or composition of matter, or any new and useful improvement thereof." Utility patents protect the structure or function of an invention for a term of 20 yearsfrom their earliest effective filing date.

Design patents protect "any new, original, and ornamental design for an article of manufacture" for a term of 14 years from their issue date. The subject matter of a design patent may relate to the configuration or shape of an article, to the surface ornamentation on an article, or to both. If a design is primarily the result of an article's function, a utility patent may be preferable over a design patent. For example, the following patents illustrate both a utility patent and a design patent for an ice-cream cone.

In these examples, the utility patent (a) protects functional aspects of the ice-cream cone, such as a "a preformed, closed-bottomed wafer shell" and "a separate, preformed, closed-bottomed chocolate shell." In contrast, the design patent (b) provides a different scope of protection directed to the appearance of the ice-cream cone shown in these examples.

Plant patents protect "asexually reproduces any distinct and new variety of plant, including cultivated sports, mutants, hybrids, and newly found seedlings, other than a tuber propagated plant or a plant found in an uncultivated state" for a term of 20 years after its earliest effective filing date. Asexual reproduction means to reproduce a plant without seeds, and

US006235324B1

(12) **United States Patent**
Luigi Grigoli et al.

(10) **Patent No.:** **US 6,235,324 B1**
(45) **Date of Patent:** **May 22, 2001**

(54) **COMPOSITE ICE-CREAM CONE**

(75) Inventors: **Franco Albino Luigi Grigoli; Ivano Maini**, both of Milan (IT)

(73) Assignee: **S.I.D.A.M. S.R.L.,** Cormano (IT)

(*) Notice: Subject to any disclaimer, the term of this patent is extended or adjusted under 35 U.S.C. 154(b) by 0 days.

(21) Appl. No.: **09/267,474**

(22) Filed: **Mar. 11, 1999**

(30) **Foreign Application Priority Data**

Jun. 1, 1998 (IT) MI98A1210

(51) **Int. Cl.⁷** A23G 9/00
(52) **U.S. Cl.** 426/90; 426/94; 426/95; 426/101; 426/138; 426/139
(58) **Field of Search** 426/95, 139, 138, 426/391, 390, 100, 101, 93, 94

(56) **References Cited**

U.S. PATENT DOCUMENTS

1,509,194	* 9/1924	Dresser	426/138
1,607,664	* 11/1926	Carpenter	426/139
1,690,984	* 11/1928	Lane et al.	426/139
1,875,960	* 9/1932	Turnbull	426/139
1,876,105	* 9/1932	Turnbull	426/139
1,938,113	* 12/1933	Schoenfeld	426/139
1,988,392	* 1/1935	Niklason	426/95
2,135,808	* 11/1938	Friedman	426/95
2,167,353	* 7/1939	Frediani	426/95
2,248,448	* 7/1941	Chester	426/139
2,527,993	* 10/1950	Habler	426/139
2,649,057	* 8/1953	Niklason	426/95
2,759,826	* 8/1956	Lindsey	426/95
4,390,553	* 6/1983	Rubenstein et al.	426/139
4,427,702	* 1/1984	Andrews	426/139
4,472,440	* 9/1984	Bank	426/138
4,600,591	* 7/1986	Galli	426/138
5,858,428	* 1/1999	Truscello et al.	426/138

FOREIGN PATENT DOCUMENTS

416891	* 9/1934	(GB)	426/139
568385	* 10/1957	(IT)	426/139
3-240442	* 10/1991	(JP)	426/101
6-278786	* 10/1994	(JP)	426/138

* cited by examiner

Primary Examiner—Steven Weinstein
(74) *Attorney, Agent, or Firm*—Notaro & Michalos P.C.

(57) **ABSTRACT**

A composite ice-cream cone having a preformed, closed-bottomed wafer shell. The shell has an inner surface, a length, an outer surface and an upper edge. A separate, preformed, closed-bottomed chocolate shell has a lower portion, an upper portion and a filling of a frozen product contained in both the lower and upper portions. The lower portion of the chocolate shell is insertedly nested within the wafer shell with the lower portion having a form complimentary to the inner surface of the wafer shell such that lower portion extends over both the entire inner surface of the wafer shell and the length of the wafer shell.

2 Claims, 2 Drawing Sheets

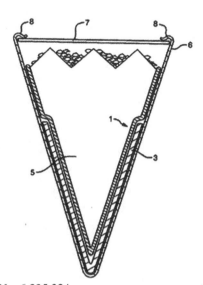

(a) U.S. Utility Patent No. 6,235,324

US00D482181S

(12) **United States Design Patent**   (10) Patent No.:   **US D482,181 S**

Carbone   (45) **Date of Patent:**   ** **Nov. 18, 2003**

(54) **ICE CREAM CONE**

(75) Inventor: **Arnold James Carbone**, Burlington, VT (US)

(73) Assignee: **B & J Homemade, Inc.**, South Burlington, VT (US)

(**) Term: **14 Years**

(21) Appl. No.: **29/150,635**

(22) Filed: **Nov. 27, 2001**

(51) LOC (7) Cl. **01-01**
(52) U.S. Cl. **D1/118**
(58) Field of Search D1/101–106, 116–130, D1/199; 99/383; 426/94, 95, 100, 101, 104, 138, 139, 144, 289, 290, 291, 292, 295, 549, 439, 496

(56)   **References Cited**

U.S. PATENT DOCUMENTS

D54,440	S	*	2/1920	McLaren D1/114
D56,488	S	*	10/1920	Winder D1/118
2,167,353	A	*	7/1939	Frediani 426/102
D131,946	S	*	4/1942	Hale D1/102
D153,352	S	*	4/1949	Halset D1/118
2,749,853	A	*	6/1956	Graham 99/383
5,916,611	A	*	6/1999	Bell 426/95

OTHER PUBLICATIONS

RolloCone on p. 1 of the Matterhorn Company Winter 1999 newsletter © 2002, retrieved from the Internet on Oct. 3, 2002 from:<URL: http://www.thematterhorn.com/matdev/newsletter_winter_1999.htm.*

* cited by examiner

Primary Examiner—Alan P. Douglas
Assistant Examiner—Linda Brooks
(74) *Attorney, Agent, or Firm*—Gerard J. McGowan, Jr.

(57)   **CLAIM**

The ornamental design for an ice cream cone, as shown and described.

DESCRIPTION

FIG. 1 is a perspective view of our new invention;

FIG. 2 is a rear elevational view of our invention;

FIG. 3 is a front elevational view of our invention;

FIG. 4 is a side elevational view of our invention;

FIG. 5 is a side elevational view of our invention taken from the opposite side;

FIG. 6 is a top plan view of our invention;

FIG. 7 is a bottom plan view of our invention;

FIG. 8 is a perspective view of a first alternate embodiment of our invention;

FIG. 9 is a rear view of a first alternate embodiment of our invention;

FIG. 10 is a front view of a first alternate embodiment of our invention;

FIG. 11 is a first side view of a first alternate embodiment of our invention;

FIG. 12 is a second side view of our invention taken from the opposite side, of the first alternate embodiment of our invention;

FIG. 13 is a top plan view of the first alternate embodiment of our invention; and,

FIG. 14 is a bottom plan view of the first alternate embodiment of our invention.

1 Claim, 4 Drawing Sheets

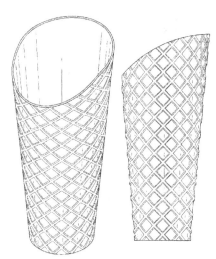

(b) U.S. Design Patent No. 482,181

includes techniques such as grafting, budding, or using cuttings, layering, or division in order to assure that offspring are substantially identical to the parent. Naturally occurring plant varieties, however, are not patentable.

1.3 Provisional Patent Applications

In advance of filing a non-provisional or regular patent application, a provisional patent application can be filed to preserve an early filing date for 1 year. The requirements for filing a provisional application include a specification, drawing figures (if necessary to an understanding of the invention), the official filing fee, and the name and home residence of each inventor. A provisional patent application is not examined by the USPTO. A provisional application can be converted into a non-provisional patent application at any time during this 12-month period. In addition, an applicant has 1 year from its provisional patent application filing date to file any foreign patent applications claiming priority to the provisional patent application filing date. The benefits of a provisional application are lower cost, right to an earlier effective filing date, and minimal filing requirements. Provisional patent applications remain confidential (and a potential trade secret) if the 12-month period lapses and the applicant decides not to pursue a non-provisional patent application.

Example: Dr. Curd inadvertently forgot to inform Tastewell that he submitted a description for Tastewell's new fruit and cheese product to the Dairy Times for publication. Dairy Times will publish tomorrow, and Tastewell wants to file a patent application to preserve its rights to file a foreign patent application before the publication.

Because Dairy Times publishes tomorrow, it is unlikely that Tastewell and its patent attorney will have sufficient time to prepare a thorough non-provisional patent application. Under this scenario, Tastewell should file a provisional application with as much data as it can possibly submit. Tastewell will then have 1 year to file a non-provisional patent application or any foreign patent applications claiming priority to its provisional application's filing date.

1.4 The Patent Application Parts

Once the decision is made to pursue patent protection for an invention, a patent application must be filed with the United States Patent and Trademark Office (USPTO). The USPTO assigns a filing date and application serial number to the application. The filing date is important because it sets a "soft" date in which "prior art" references must predate in order to reject the claims of the application.

A regular or non-provisional patent application must provide a specification that: (1) describes the invention in sufficient detail to show one skilled in the art that the inventor possessed the claimed invention at the time of filing ("written description requirement"); (2) describes the invention in a manner that would allow one skilled in the art to make and use the claimed invention without undue experimentation (e.g., "enablement requirement"); and (3) discloses the preferred way of carrying out the claimed invention at the time the application is filed ("best mode requirement"). The specification must conclude with one or more claims that particularly point out and distinctly claim the novel subject matter of the invention. Drawings are "necessary for the understanding of the subject matter sought to be patented."[1] Finally, the application must include an oath or declaration naming the true and correct inventors, and must include the requisite filing fee.

1.4.1 Claims

A patent application's claims are critical to defining the scope of protection sought in a patent. The claim's scope has been described as defining the "metes and bounds" of the patented invention. These "metes and bounds" define a patent holder's rights to exclude others from making, using, selling, offering to sell, or importing an accused invention. Thus, if an accused invention falls within a patent claim's scope, it infringes the patent's scope. Before reaching infringement, however, the claims must meet certain requirements.

The claims must be supported by the specification. If the specification describes parts of an invention that are not defined in the claims, it is possible that such disclosure will be dedicated to the public. For this reason, the claims must particularly point out and distinctly claim the novel subject matter of the invention, and should describe the invention as broadly as possible based on the specification.

The claims in a patent application are typically structured to include independent claims that broadly define the claimed invention, and dependent claims that limit the scope of the independent claims. A dependent claim includes all of the limitations of an independent claim, and also includes additional elements that further limit the independent claim. For example,

---

[1]37 C.F.R. § 1.81(a).

a dependent claim may read, "The apparatus of claim 1, further comprising [additional elements]."

> *Example*: Suppose that Tastewell's mixed cheese and fruit product is described in its patent application specification as being a mixture of processed cheese with any type of fruit. However, Tastewell made a decision prior to filing its patent application that it will only sell the cheese mixed with strawberries, and only pursues patent claims to the cheese and strawberry embodiment. Five years after Tastewell's patent issues, Tastewell realizes that consumers would prefer a variety of cheese and fruit combinations, and Tastewell wants to pursue patent protection to cover other embodiments.
>
> Because Tastewell disclosed, but did not claim, a mixture of cheese with any type of fruit in its original patent application, it is now likely barred from attempting to obtain patent protection for other fruit embodiments. Furthermore, by not claiming this subject matter, Tastewell has dedicated it to the public. Tastewell could have attempted to file a broad independent claim in its original patent application as a mixture of cheese with any type of fruit and sought more narrow protection to specific fruits in dependent claims.

1.4.2 Specification

1.4.2.1 Written Description Requirement

The specification must fully describe the invention recited in the claims with particularity. While the specification does not have to describe the claims verbatim, it must describe the claimed invention in such a manner that a person of ordinary skill in the application's technical field would understand the claimed invention. In addition, the specification should describe as many alternate embodiments of the invention as reasonably permissible in order to avoid any rejections by the USPTO for lack of written description.

> *Example*: Tastewell's patent application specification describes a mixture of processed cheese combined with any type of fruit. After filing its patent application, Tastewell learns that consumers prefer a mixture of cheese and vegetables, and would like to pursue patent protection for this embodiment.

Because Tastewell's original patent application did not disclose a mixture of cheese with vegetables, any claims to such an embodiment will likely be rejected as not being supported by the specification.

1.4.2.2 Enablement Requirement

The enablement requirement requires that the specification, at the time the application is filed, describe the invention in such a manner that a person of ordinary skill in the art could *make* and *use* the claimed invention without undue experimentation. The fact that a person of ordinary skill in the art is required to perform *some* experimentation when carrying out the claimed invention does not mean that such experimentation is "undue." However, the quality or quantity of any such experimentation must not be unreasonable or unduly burdensome.

1.4.2.3 Best Mode Requirement

To satisfy the best mode requirement, the patent application, at the time of filing, must describe the inventor's preferred way of carrying out the claimed invention. This requirement is subjective because it is dependent on the inventor's state of mind, and not necessarily on whether the description provides the true best mode of carrying out the invention. The best mode requirement is intended to prevent the inventor(s) from concealing a preferred embodiment of the invention from the public. Thus, in order to fail to satisfy the best mode requirement, the inventor must know of a better way of carrying out the claimed invention and conceal it at the time of the application's filing.

1.4.3 Inventorship

A US patent application must be filed by the actual inventor(s) of the subject matter. Determination of inventorship can be a difficult task that requires legal analysis. "Conception" of the invention is typically considered the key for determining inventorship. Conception is the mental formulation and disclosure by the inventor or inventors of a complete idea for a product or process. Contributions of labor or supervision are typically insufficient to vest inventorship rights in the invention. In contrast, in the academic setting, it is often discretionary to name contributors of a research project on published articles. However, naming inventors of a patent application is not discretionary. If the inventorship on an issued patent is incorrect, a court can invalidate the patent.

> *Example*: Two of Tastewell's senior scientists, Dr. Curd and Dr. Whey, equally contributed to the conception of the cheese and fruit product mixture. Two entry level scientists, Dr. Apple and Dr. Orange, initially tested the product to determine the ratio and amounts of ingredients under the direction of Dr. Curd. After Drs. Curd and Whey accepted the product, entry level scientist Dr. Orange discovered that the product has a longer shelf life by increasing the heat exposure only during processing. Tastewell decides to file a patent application for this invention and needs to determine the inventor(s).
>
> In this example, if Tastewell decides to pursue claims in a patent application directed to the composition of the new cheese product, Dr. Curd and Dr. Whey should be considered the inventors. If Tastewell decides to pursue claims to a method of making the product that involves increased heat exposure to increase shelf life then Dr. Curd, Dr. Whey, and scientist Dr. Orange should be identified as the inventive entity.

1.4.4 When Should You Apply for a Patent Application?

Currently, the USPTO grants a US patent to the first inventor to invent. Most other countries rely on a first inventor to apply (first-to-file) system. In either case, when pursuing patent protection, an inventor should file a patent application as soon as the invention is complete. An invention is considered to be complete after it is conceived *and* reduced to practice. Conception is an inventor's mental formulation and disclosure of a complete idea for a product or process. The test for conception is whether the inventor had an idea that was definite and permanent enough that a person skilled in the art could understand the invention. Completion of an invention's second requirement, reduction to practice, has two types: "constructive" and "actual." Filing a complete patent application satisfies a "constructive" reduction to practice. To prove "actual" reduction to practice, an inventor must have: (1) constructed an embodiment of the invention; and (2) tested the device or process so as to establish its capacity to successfully perform its intended purpose.

After conception, it is important that the inventor act diligently to reduce an invention to practice and file a patent application with the USPTO. The

USPTO gives the application a "soft" date (i.e., the application filing date) for determining the date that any "prior art references" must precede in order to be cited against the claims in the application. A prior art reference is anything that was publicly available prior to the date of invention. The application filing date is also useful for establishing a priority date over other similar or competing patent applications. The filing date is not, however, the final arbiter of which of two competing applications is entitled to a patent. For example, it may be possible to "swear behind" the application filing date by showing an earlier date of actual reduction to practice, or an early date of conception coupled with diligent reduction to practice. Similarly, during prosecution of a patent application, an applicant can similarly "swear behind" a prior art reference.

Example: Dr. Curd and Dr. Whey conceived Tastewell's new cheese and fruit product on January 1, 2000. Shortly after Dr. Curd and Dr. Whey conceived Tastewell's new cheese and fruit product, Dr. Curd resigned from Tastewell Industries, and began working for Tastewell's competitor, Bland Foods. Dr. Whey completed a working embodiment of the cheese and fruit product on May 1, 2000, and filed a patent application for the product on June 1, 2000. Tastewell learned that its competitor, Bland Foods, began marketing a similar cheese and fruit product shortly after Dr. Curd was hired. Bland Foods reduced its product to practice on March 1, 2000, and filed a patent application for the product on April 1, 2000. Tastewell wants to know whether it can claim priority over Bland Foods patent application.

Timeline of Events

Tastewell's conception	Bland Foods' reduction to practice	Bland Foods' application filing date	Tastewell's reduction to practice	Tastewell's application filing date
01/01/2000	03/01/2000	04/01/2000	05/01/2000	06/01/2000

Because Bland Foods reduced its invention to practice prior to Tastewell, Bland Foods is considered to have priority over Tastewell. However, if Tastewell can provide sufficient evidence, such as an inventor's notebook, that proves it acted diligently to reduce its invention to practice between its January 1, 2000 date of conception and its May 1, 2000 date of reduction to practice, it may be able to claim priority over Bland Foods patent application.

1.4.5 Patent Examination

After filing a patent application, the USPTO assigns the application to a patent examiner for examination. During examination, the examiner ensures that the application satisfies all formal requirements for the specification, claims, and drawings. The examiner also conducts a search of available prior art references using search databases, including the Internet. Following the examiner's initial examination and search, the examiner will usually issue an objection to the application for failing to satisfy a formal requirement, or reject the claims as anticipated or obvious in view of the prior art discovered during the examiner's search.

In response to an Official Action, the applicant, typically through his or her attorney, can submit a formal response to address the rejections noted by the examiner and distinguish the claimed invention over the prior art. By distinguishing the claimed invention over the prior art, the applicant may amend the claims. Claim amendments are not required and may be particularly unnecessary when an examiner misinterprets a reference or improperly combines references to support a rejection.

After filing a response to an Official Action, the examiner considers the arguments or amendments and makes a determination as to whether to issue a subsequent Official Action to allow the application. If the examiner issues another Official Action, the Applicant will be given opportunity to respond. There is no limit on the number of Official Actions that can be issued in a patent application, although after a first action, examiners will usually issue a final office action which can have the effect of closing prosecution. If prosecution is closed in an application, an applicant can file a Request for Continued Examination (RCE) along with a response and the USPTO official fee. If the examiner decides to allow the application, the applicant will receive a Notice of Allowance, which will have a set period for the applicant to pay a fee in order to have the application officially issued as a patent.

1.4.6 Continuing Applications

US patent law allows applicants to file continuing patent applications claiming the benefit of the disclosure and filing date of an earlier filed co-pending application ("parent application"). The parent application does not have to be the first or earliest filed application in a chain of continuing applications; it just has to be a related application that is co-pending at the time of filing. Continuing applications must share at least one common inventor with the parent application, make a specific claim of priority to the parent application, and be filed while the parent application is co-pending. Although continuing applications claim the benefit of the earlier filed "parent" application, they are newly filed applications that restart the examination process.

There are three types of continuing patent applications recognized in US patent practice: (1) continuation applications; (2) continuation-in-part applications; and (3) divisional applications.

Continuation applications have the same specification as the parent application but with different claims. Continuation applications are useful to: (1) claim subject matter that was disclosed but not fully claimed in the parent application; (2) seek broader or different claim coverage; (3) present new arguments in support of allowance of the application after a final rejection is received or prosecution is closed;[2] or (4) keep an application pending to capture developments not specifically addressed by any of the issued claims.

Divisional applications can be filed in response to an Office Action from the USPTO which states that the claims of the parent application are directed to two or more distinct inventions (e.g., claims to a product and claims to a method of making a product can be considered distinct inventions).

A continuation-in-part application (CIP) is a later filed application that repeats some substantial portion, if not all, of the parent application's disclosure, and, generally, adds new subject matter not disclosed in the parent application. Claimed subject matter that is supported by the parent application is entitled to the effective filing date of the parent application. Claimed subject matter that is not supported by the parent application has the filing date of the CIP. Generally, CIPs claim new or related embodiments of an invention not disclosed in the parent application while effectively maintaining the filing date of the parent application for all originally disclosed subject matter. Although the applicant always has the option of filing a new application for this new subject matter, the priority claim for a CIP application may prevent the parent application itself from being cited to reject any original subject matter from the parent application that is claimed in the CIP.

Example: Tastewell's patent application describes its mixed cheese and fruit as being a mixture of processed cheese with any type of fruit. After filing its patent application, Tastewell learns that consumers

---

[2] A Request for Continued Examination (RCE) under 37 C.F.R. § 1.114 can be filed upon payment of the requisite fee to present new arguments or claims in an application after a final rejection as an alternative to filing a continuation application.

prefer a mixture of cheese and vegetables, and would like to pursue patent protection for this embodiment.

Assuming that the cheese and fruit product and cheese and vegetable product have common compositions, Tastewell may be able to file a continuation-in-part (CIP) application to describe and claim the new embodiment to a cheese and vegetable product. By filing a CIP application, the priority date for all subject matters in the CIP that overlap with the original application disclosure will have the original application's filing date. All new subject matters will be entitled to the CIP application's filing date. Such a continuation-in-part application must be filed while Tastewell's original patent application or an application claiming priority thereto is still pending (i.e., not an issued patent).

1.5 Patentability Requirements

In order for something to be patentable, it must be (1) patentable subject matter; (2) useful; (3) novel; and (4) non-obvious.

1.5.1 Patentable Subject Matter

Pursuant to the patent statute, "[w]hoever invents or discovers any new and useful process, machine, manufacture, or composition of matter, or any new and useful improvement thereof, may obtain a patent therefore, subject to the conditions and requirements of this title."[3]

A "process" is a way to produce a result. A process can consist of mixing cheese with fruit at a certain temperature. Not all processes are patentable. For example, a pure mathematical algorithm is not patentable. However, a mathematical algorithm included in a process used to determine a useful, concrete, and tangible result will in most circumstances be considered patentable subject matter. A "machine" is a device with assembled parts that move to perform a desired operation. A "manufacture" or "article of manufacture" is typically regarded as a man-made, tangible object that is not naturally occurring. A "composition of matter" is any compound, substance, mixture, etc. that is the result of combining two or more ingredients.

Based on theabove definitions, it is no surprise that patentable subject matter has been said to "include anything under the sun that is made by

---

[3]35 USC § 101

man."[4] There are, however, some recognized exceptions including: (1) laws of nature (2) natural phenomena; and (3) abstract ideas.

Inventions may often encompass more than one category of patentable subject matter. Accordingly, patents will often have more than one type of claim.

> *Example*:   During product testing of Tastewell's new cheese and fruit product, Tastewell discovers that the product has a longer shelf life when it increases the heat exposure during processing. Tastewell decides to file a patent application.
>
> In this example, Tastewell may be able to pursue protection for both the product and the method of making the product. A patent with product claims may give Tastewell broader protection because they would give Tastewell the right to exclude competitors from making the product according to any method. Method claims, however, are often desirable because even if the product is not held to be novel, the method of making the product can still be novel.

1.5.2 Utility Requirement

A patent application must also demonstrate that the claimed invention is "useful" for some purpose to meet the utility requirement. In most technical fields, this utility requirement has a low threshold which is easily satisfied by demonstrating any useful result. For a patented invention to fail to satisfy the utility requirement it must be "totally incapable of achieving a useful result," which is rare in applications for processes, machines, and articles of manufacture.

While rare in those instances, failure to satisfy the utility requirement is more common in biotechnology and chemical applications. In biotechnology and chemical fields, the USPTO typically requires that applications disclose a practical or real-world benefit available from the invention; in other words, a specific, substantial, and credible utility. Specific utility requires that the applicants have knowledge of what the invention does. Credible utility requires that the claimed invention be believable based on current state of the art. Finally, substantial utility requires that the claimed invention has a real-world benefit (e.g., a treatment for a disease). In the chemical field,

---

[4] *Diamond v. Chakrabarty*, 447 U.S. 303 (1980).

claims may be rejected for lack of utility if a compound or reaction creates a reasonable doubt as to whether there is a credible utility.

1.5.3 Novelty Requirement

In order for an invention to be patentable, it must be new or "novel" (i.e., not in the prior art). If the prior art shows every element of a claim, the claim is unpatentable as "anticipated" by the prior art. In the United States, prior art is "everything" in the public domain that existed before the date of "invention" or 1 year prior to the filing date of a patent application. In order for a patented invention to be rejected over a prior art reference, the reference must have been accessible somewhere in the world. Secret or non-public materials cannot act as prior art. Rules regarding prior art differ around the world. For most foreign countries, prior art is "everything" prior to the priority filing date of a patent application (i.e., most countries do not recognize a "1 year" grace period).

1.5.3.1 Prior Invention: 35 USC § 102(a)

If an invention was known or used by another in the United States or in a printed publication anywhere in the world before the *date of invention*, it is not patentable. Thus, the scope of prior art includes any printed publication that predates the date of invention and discloses each and every element of a patent claim. US patent law looks to the date of invention and not the patent application's filing date. Therefore, if the USPTO cites prior art that predates an application, it is possible to "swear behind" a reference by showing an earlier date of invention. This procedure is unique to the United States. Nearly all other patent systems define the date of invention as the application's filing date.

Section 102(a) also considers knowledge or use by another in the United States. The phrase "by another" means any person other than the inventive entity for the patent application. For example, displaying a product at a trade show is typically considered a use that bars patentability.

Example: Tastewell reduces the cheese and fruit product to practice on January 1, 2000, and files a patent application on March 1, 2000. Tastewell's rival, Bland Foods, publishes an article disclosing the same product on February 1, 2000. During examination of Tastewell's patent application, a USPTO examiner relies on Bland Foods publication as anticipating Tastewell's patent application's claims.

In order to overcome this reference, Tastewell can submit evidence establishing a date of invention (i.e., January 1, 2000) that predates Bland Foods publication date. The evidence, however, must be sufficient to prove the date. Sufficient evidence might include an inventor's notebook or dated prototype. The sufficiency of any such evidence depends on Tastewell and its inventors' record-keeping practices. In most foreign jurisdictions, however, the Bland Foods' article will be treated as prior art.

1.5.3.2 Statutory Bars: 35 USC § 102

Section 102(b) is often referred to as the "statutory bar" provision. Pursuant to this section, disclosure of an invention, anywhere in the world, more than 1 year before applying for a patent is a bar to obtaining a patent. Any public disclosure, use, offer for sale, or sale of the claimed invention by another, with or without the inventor's consent, can be a statutory bar. Public disclosure of an incomplete invention may not rise to a statutory bar.

Section 102(b) statutory bar can often be the result of the inventor's own actions. For example, an inventor's public disclosure of the invention at a trade show or offer to sell the invention to anyone more than 1 year prior to the application filing date can be a statutory bar. An exception to the public use statutory bar is if the invention is being publicly used for bona fide testing or evaluation.

Example: Dr. Curd and Dr. Whey equally contribute to making the new cheese and fruit product. Dr. Whey is interested in publishing an article for the food industry to disclose their new breakthrough cheese product.

If Tastewell intends to pursue patent protection for its new product, Dr. Curd should wait until after the patent application is filed to publish his article. If Dr. Curd publishes his article prior to the date the application is filed, Tastewell will have 1 year from the first date of circulation of the publication to file its US patent application. By publishing the article prior to Tastewell's application filing date, Tastewell may be prohibited from filing foreign patent applications.

Example: During the research and development phase of the cheese and fruit product, Tastewell wants to test children's allergenic reactions to the product to determine whether the product ingredients need to be modified. In doing so, Tastewell goes to the local recreation center and allows children to test the product under the condition that they agree to undergo evaluation. After it completes this testing and makes a determination that the product is acceptable, Tastewell makes a competitor's product available to children at the recreation center for a 1 day only "taste-testing" in order to evaluate their preference. One year and 1 day after this taste-testing, Tastewell files a patent application for the product.

In the above example, Tastewell will likely be able to argue that the allergenic testing is not a statutory bar because it was conducted for bona fide experimental purposes in order to determine whether the product composition is acceptable. In contrast, the "taste-testing" evaluation will likely create a statutory bar that prevents patentability of the product because evaluating consumer preference is typically not considered an experimental purpose.

The following chart summarizes the types of materials and acts that are considered "prior art":

What	Who	Where	When
The invention is publicly known	By another	In the United States	Before the applicant's date of invention
The invention is publicly used	By another	In the United States	Before the applicant's date of invention
The invention is described in a patent	By another	Anywhere	Before the applicant's date of invention
The invention is described in a publicly available printed publication	By another	Anywhere	Before the applicant's date of invention
The invention is described in a patent	By anyone	Anywhere	More than 1 year prior to the application filing date

What	Who	Where	When
The invention is described in a publicly available printed publication	By anyone	Anywhere	More than 1 year prior to the application filing date
The invention is publicly known	By anyone	In the United States	More than 1 year prior to the application filing date
The invention is publicly used	By anyone	In the United States	More than 1 year prior to the application filing date
The invention is on sale	By anyone	In the United States	More than 1 year prior to the application filing date

1.5.3.3 Non-Obviousness Requirements: 35 USC § 103

An invention is obvious if the differences between the subject matter sought to be patented and the prior art are such that the subject matter as a whole would have been obvious at the time the invention was made to a person having ordinary skill in the art to which said subject matter pertains. For example, merely using a screw for a nail would normally not be patentable, since both are commonly used fasteners.

In conducting an obviousness analysis, an examiner may combine multiple prior art references. The examiner cannot, however, combine references arbitrarily. The non-obviousness requirement requires that an examiner step into the shoes of a person of ordinary skill at the time the invention was made and determine whether the claimed invention would have been obvious without using hindsight obtained by reviewing the patent application.

Every obviousness determination considers four factual inquiries: (1) the scope and content of the prior art; (2) the differences between the prior art and the claimed invention; (3) the level of ordinary skill in the pertinent art field at the time of the invention; and (4) objective evidence of obviousness or non-obviousness ("secondary considerations").

The scope and content of the prior art includes art that is directed to the same field of invention as claimed in a patent application, and any other art that is logically relied upon. The prior art used in determining whether an invention is obvious is the same material defined as "prior art" under

35 USC § 102. Using the above example, if an invention is directed to a cheese product including slices of fruit, an examiner might look to the cheese art, fruit art, yoghurt art, and any other art concerned with combining cheese or fruit with another substance.

Determining the differences between the prior art and the claimed invention is a useful starting point to determine whether the claimed invention would have been obvious in view of the prior art. If the differences between the prior art and claims are trivial, the claimed invention will likely be unpatentable as obvious in view of the prior art.

The level of skill required of a hypothetical person having ordinary skill in the art is more than an ordinary layperson but less than an expert in the field of the invention. Determining the level of skill in the art is a factual question that is often open to debate. Factors that are often considered in such a determination can include the level of sophistication in the technology, the education of ordinary person in the field, and prior art attempts to solve related problems.

Courts refer to objective evidence of obviousness or non-obviousness as "secondary considerations." Such secondary considerations include: long felt need for the invention, commercial success of the invention, and copying by others. For example, if there was a long need for the claimed solution to a problem, or if the invention is commercially successful, the claimed invention is likely not obvious. Also, showing the prior art teaches away from the claimed invention can be used to support non-obviousness.

1.6 International Patent Rights

Rules for obtaining a patent differ from country to country. Patent protection in other countries requires international filings, usually with each country's patent office. Most countries permit applicants a non-extendible period of 1 year from the date of filing a US patent application in which to file their patent application. In most countries, if a foreign patent application is filed within this 1 year period and claims priority to a US patent application, the US patent application filing date is the applicable priority date of the application.

The United States and approximately 120 other countries abide by the Patent Cooperation Treaty that permits patent applicants to file international patent applications, also know as PCT applications. A PCT application is similar to a US provisional application in that it preserves priority and never issues a patent. Within 30 months from the PCT priority date, the applicant must file individual patent applications in all countries in which examination is desired (i.e., PCT applications provide an additional 18 months time to file foreign application beyond the typical 1 year period for filing priority foreign applications). Filing a PCT application can be advantageous in the following respects:

(1) if an applicant is interested in filing a patent application in numerous countries, a PCT application permits the applicant to have the benefit of a PCT patent examiner's prior art search and results before incurring the expense of filing numerous patent applications;
(2) a PCT application gives an applicant additional time (30 months from the PCT filing date) to delay the expenses associated with applying for patent protection in individual countries; and
(3) many countries give credence to a PCT examiner's examination search and opinion on patentability, which can limit the costs of prosecuting a patent application in individual countries.

Example: Tastewell wants to pursue US and foreign patent protection for its cheese and fruit product. Tastewell would like to file its patent applications as soon as possible, but is unsure as to how successful the product will be and is hesitant to spend too much on international patent protection.

If Tastewell files a PCT application, it will have up to 30 months to determine in which countries to pursue protection. This will give Tastewell additional time to evaluate the commercial success of the product and target select foreign markets. A PCT application will also give Tastewell the benefit of a single examination, which can assist its determination of how much to invest in both United States and international patent protection.

Chapter 2
Trade Secret Protection

Trade secret law provides a mechanism for protecting proprietary and sensitive business information. A trade secret, by definition, is information that has economic value and is secret. There are no formal application requirements to obtain a trade secret. Unlike patents, there are no statutory requirements that a trade secret be novel, useful, non-obvious, and there is no examination process. Trade secret protection arises once the appropriate steps are taken to create a valid trade secret. Trade secrets are not subject to a predefined term, and can be maintained for an indefinite period of time.

2.1 What Is a Trade Secret?

Unlike patent law, which has its roots firmly grounded in federal constitutional and statutory law, trade secret law is a state law doctrine that developed out of the common law doctrine of unfair competition and unfair business practices. Until passage of the Uniform Trade Secrets Act (UTSA) in 1985, trade secret law varied significantly from state to state. The UTSA is a model law that provides a uniform definition of trade secrets and misappropriation, and 45 states, the US Virgin Islands, and the District of Columbia, have adopted it.

The UTSA defines a trade secret as "information, including a formula, pattern, compilation, program, device, method, technique, or process, that: (1) derives independent economic value, actual or potential, from no being generally known to, and not being readily ascertainable by proper means by, other persons who can obtain economic value from its disclosure or use and (2) is the subject of efforts that are reasonable under the circumstances to maintain its secrecy." This broad definition maintains the common law that nearly any type of business information can qualify as a trade secret. Thus,

R.W. O'Donnell et al., *Intellectual Property in the Food Technology Industry*,
DOI: 10.1007/978-0-387-77389-6_2, © Springer Science+Business Media, LLC 2008

information that is not otherwise patentable can be a trade secret. Examples
of information that can be protected by trade secret include:

- computer programs
- client identities
- product pricing
- manufacturing processes
- technical information
- technical information
- prototypes
- company manuals
- financial statements
- customer lists
- vendors
- market analysis and strategies
- formulas
- product testing results (positive and negative)
- drawings
- strategic plans
- employee records and salaries
- product ingredients (foods, cosmetics, or drugs, etc.)

Because information of nearly any type of subject matter can qualify as a
trade secret, the UTSA definition of a trade secret focuses on: (1) the economic value of the trade secret; (2) whether the trade secret is generally
known or readily ascertainable; and (3) the efforts taken to maintain secrecy.
The "economic value" requirement under the UTSA refers to whether a
competitor would obtain an economic benefit if the trade secret information
became readily accessible. "Economic value" can be shown by the time and
effort utilized in creating the trade secret, or by showing that a third party
would have to spend time and effort in creating the same trade secret.

The second requirement for a trade secret under the UTSA is that the
information cannot be "generally known or readily ascertainable." This
means that the information cannot be already known to the public or by competitors. Whether a trade secret is "generally known or readily ascertainable"
is a factual inquiry that depends on the amount of time, effort, and money
required to independently produce the trade secret, or to reverse engineer the
trade secret. Information cannot be protected by a trade secret if it can be
discovered by examining a commercially available product that incorporates
the information. If the trade secret is hidden in a commercially available
product, then the trade secret can be maintained. A trade secret that consists of the amounts and ratios of individual ingredients in a product or code

embedded in a software program is not lost just because the product becomes public availability.

Published information, such as that disclosed in a book, magazine, trade publication, website, or other media, cannot be maintained as a trade secret because it is "generally known" and readily ascertainable. This can be particularly important when deciding whether to keep information as trade secret or to pursue patent protection for that information. Anything disclosed in a patent or published patent application is generally known and readily ascertainable and cannot be protected as a trade secret.

Example: Tastewell is confident that the ratio of ingredients in its new cheese and fruit product could not be reverse engineered by a competitor analyzing its product. However, as part of a marketing strategy, Tastewell decides to pursue patent protection for the formulation of its cheese and fruit product. During examination of its patent application, the USPTO examiner asserts that Tastewell's formulation would be obvious. Tastewell is unable to convince the examiner otherwise and decides to abandon its patent application. Tastewell inquires whether its product formulation can be maintained as a trade secret now that it cannot get a patent.

If Tastewell's patent application was not published before abandonment, it may be able to maintain its production formulation as a trade secret. However, if Tastewell's patent application is published, the information is public and Tastewell cannot maintain its product formulation as a trade secret.

The final and often most important criterion for a trade secret under the UTSA is that reasonable efforts must be taken to maintain secrecy of the information. Maintaining secrecy of a trade secret is viewed under a reasonable standard which does not require absolute secrecy. A court considers several factual inquiries when considering reasonable secrecy:

- whether employees have executed confidentiality or non-disclosure agreements;
- whether the company's confidentiality policy is memorialized in writing;
- whether access to the trade secret is been limited to essential employees/contractors;
- whether employees who are privy to the trade secret are aware that it is to be maintained as a trade secret;

- whether the information is kept in a restricted area such as a locked file, within security encrypted software, in a restricted location within a physical plant, etc.;
- whether documents containing information that is trade secret are properly labeled; and
- whether the company actively screens employee publications, presentations, etc. for disclosure of trade secret information.

In addition to these factors, it is important that the owner of the trade secret takes steps to enforce secrecy of the information. Mere intent to keep information trade secret, without affirmative acts, is typically insufficient to maintain a trade secret.

2.2 Misappropriation of Trade Secrets

A trade secret owner has the right to prevent others from misappropriating the trade secret. The UTSA defines misappropriation of a trade secret as:

(i) acquisition of a trade secret of another by a person who knows or has reason to know that the trade secret was acquired by improper means; or
(ii) disclosure or use of a trade secret of another without express or implied consent by a person who

    (A) used improper means to acquire knowledge of the trade secret; or
    (B) at the time of disclosure or use knew or had reason to know that his knowledge of the trade secret was

        (I) derived from or through a person who has utilized improper means to acquire it;
      (II) acquired under circumstances giving rise to a duty to maintain its secrecy or limit its use; or
     (III) derived from or through a person who owed a duty to the person seeking relief to maintain its secrecy or limit its use; or

    (C) before a material change of his position, knew or had reason to know that it was a trade secret ad that knowledge of it had been acquired by accident or mistake.

In summary, misappropriation is the improper acquisition, disclosure, or use of a trade secret. A trade secret can be misappropriated even if the misappropriating party is not identically duplicating the trade secret.

Trade secrets can be lost or stolen in a variety of ways. Theft, bribery, misrepresentation, or breach of a duty to maintain secrecy are common acts that trigger a trade secret loss. Violating a confidentiality or non-disclosure agreement or obtaining the trade secret from a third party that is bound by a duty of confidentiality can give rise to an action for misappropriation. For example, a common means by which trade secrets can be lost or stolen is typically through unhappy or former employees who use or disclose the trade secret information apart from the company.

When a company discloses its trade secret to others, such as employees, manufacturers, suppliers, consultants, etc., those disclosures should be made under a written duty of confidentiality. This is typically done by requiring the party to execute a confidentiality or non-disclosure agreement, by way of employment contract, or third party consulting or supplier agreement. If a party under a duty of confidentiality with the trade secret owner breaches that duty, the trade secret owner's enforcement effort will benefit from a written agreement that clearly recognizes the trade secret status of the information.

The UTSA identifies a number of remedies for misappropriation of trade secrets including injunctions, damages, and attorney's fees. The UTSA even permits recovery of both the actual loss created by the misappropriation and any unjust enrichment resulting from the misappropriation that is not included in the "actual loss" portion of the damages. If actual loss for the misappropriation is difficult to prove, the trade secret owner may seek a "reasonable royalty" as compensation for the misappropriation. If the acts resulting in the trade secret misappropriation are willful or malicious, the UTSA grants the court discretion to award attorney's fees to the trade secret owner.

Example: Two of Tastewell's scientists, Dr. Curd and Dr. Whey, invented Tastewell's new cheese and fruit product. Subsequently, Dr. Curd resigned from Tastewell and began working for Tastewell's competitor, Bland Foods. Tastewell learned that Bland Foods began marketing a similar product almost immediately after Dr. Curd was hired. Can Tastewell take any action against Bland Foods?

Tastewell should consider an action against Bland Foods for misappropriation of trade secrets. In order to prevail, Tastewell must first establish that the cheese and fruit product was maintained as a trade secret (independent economic value, reasonable measures to maintain secrecy, and not readily ascertainable to others by proper means) and that Dr. Curd improperly disclosed it to Bland Foods. The easiest way to prove knowledge and improper disclosure is to show that

Dr. Curd had acknowledged in writing his duty to maintain Tastewell's product in secrecy (i.e., by way of a confidentiality of non-disclosure agreement) and Tastewell had written internal procedures directed to preventing disclosure.

2.3 Reverse Engineering of Trade Secrets

Under patent law, a subsequent inventor can be liable even though the invention was developed completely independently and without knowledge of the patented invention. Under trade secret law, independent discovery and use of the trade secret is not a violation. Further, competitors often try to uncover and trade off of one another's trade secrets by "reverse engineering" the trade secret; a legally acceptable practice. The comments to the UTSA state that "reverse engineering" is a proper means of discovering a trade secret and identify reverse engineering as "starting with the known product and working backward to find the method by which it was developed. The acquisition of the known product must, of course, also be by a fair and honest means, such as purchase of the item on the open market for reverse engineering to be lawful . . . ." Thus, discovery of another's trade secret requires proper acquisition of the information and ethical business practices.

Example: Once Tastewell sells its new cheese and fruit product to the public, it is permissible for any purchaser to analyze the product to determine the process by which it was produced or to determine its constituent ingredients. The purchaser has the right to use and disclose any information acquired as result of reverse engineering the product.

Chapter 3
Trademarks and Trade Dress

3.1 What Is a Trademark?

Use of symbols or signatures to identify the source of goods has been around ever since people first started trading and selling goods such as pottery, weapons, and clothing thousands of years ago. The purpose of these marks, to indicate the product's source, has not changed to this day. What has changed, especially in the last 100 years, is the protection afforded to trademarks. Currently, the United States protects trademarks under the Trademark or Lanham Act, state law, and common law.

Under the federal Lanham Act, a trademark is any word, name, symbol, device, or any combination thereof that is used to identify and distinguish goods of one source from those of another source. In short, a trademark indicates the source of the goods. The law also provides protection for other types of marks that are directed to different types of uses. Many of these different types of marks are common in the food industry, and the below table notes some of the key features of these different types of marks.

Type of mark	Key features	Example
Service mark	Used to identify and distinguish the source of services	*McDonald's*
House mark	A "house mark" generally refers to a trademark that is used in all facets of a companies' business, including business cards, letter head, packaging, and advertising. Typically, a house mark is also used with a secondary mark or can be used as a primary trademark	*Kellogg's*

R.W. O'Donnell et al., *Intellectual Property in the Food Technology Industry*,
DOI: 10.1007/978-0-387-77389-6_3, © Springer Science+Business Media, LLC 2008

Type of mark	Key features	Example
Trade dress	Trade dress refers to the overall impression created by a product which can comprise any combination of shape, color, design, and wording. If trade dress is functional it cannot be registered or protected. Product design trade dress is not registerable until there is secondary meaning	*Coca-Cola* bottle
Collective mark	Service mark used by the members of a cooperative, an association, or other collective group or organization which indicates membership in a union, an association, or other organization	*The American Institute of Wine & Food* indicating membership in an organization to promote the appreciation, understanding, and accessibility of cooking, food, drink, wine, nutrition, gardening, and agriculture
Certification mark	Mark used to certify regional or other geographic origin, material, mode of manufacture, quality, accuracy, or other characteristics of someone's goods or services, or that the work or labor on the goods or services was performed by members of a union or other organization	*PARMIGIANO REGGIANO EXPORT* for cheese's that originate in the Parma-Reggio region of Italy
Trade names	Used to identify a business or vocation. Trade names that merely identify a business are not registerable under the Lanham Act for federal registration. A trade name can also be a trademark if *used* as a trademark to indicate source. For example, Ford Motor Company can be both a trademark and a trade name	*Campbell Soup Company*

It is common for people to lump all of the above terms together as trademarks or brands. For the purpose of this discussion, the terms trademark and brand are used interchangeably herein.

A trademark normally consists of a word, logo, or some combination of the two. A word mark can include known terms, abbreviations, something coined by the owner, or some combination of letters and numbers. A logo can be a design, stylized lettering, or a drawing of an object. However, there are other types of trademarks, including the following:

- Symbol (McDonald's Arches);
- Shape (Hershey's Chocolate Kiss);
- Slogan ("*Just Do It*");
- Sound (NBC chime);
- Color (Pink for Owens Corning's insulation).
- Scents (Life-Saver candy);
- Touch ("velvet textured" feel on bottle surface for wines);
- Distinctive Packaging (T-shirt shaped box)
- Building Design (Shoney's Restaurant).

3.2 Brand Selection and Development

In the food industry, selecting the right trademark for a product can mean the difference between success and failure. Successful brand management balances legal and business considerations. The primary legal consideration is the selection of the strongest trademark possible. When considering the strength of a trademark, trademarks are ranked on a sliding scale of distinctiveness ranging from unprotectable to extremely protectable.

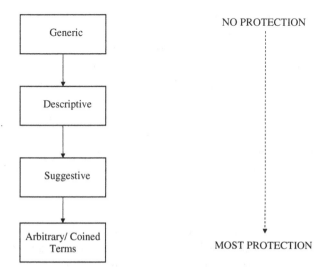

At one end of the scale are generic words. A generic word is a word that has come to be known as the common term for a class of goods or services. Generic designations are not registerable or protectable because they are incapable of functioning as a source indicator. A word can be inherently generic such as the word "cereal." Alternatively, a word can be "genericized" when the public associates the brand name as the product rather than the source. One example is "shredded wheat" for a type of cereal that is actually shredded wheat.

Next on the scale of protection are "merely descriptive" words that simply describe the product or convey an immediate idea of what the product does. One example is *OATNUT* for bread containing oats and nuts. Because descriptive marks simply describe a product, they are not protectable until they have acquired distinctiveness in the marketplace through their use. In other words, the trademark owner must show that a consumer primarily associates the mark with the product. In most cases, such distinctiveness takes years to acquire. The mark *"CALIFORNIA COOLER"* for a wine-based beverage is an excellent example of a descriptive mark that became protectable after acquiring distinctiveness. In addition, non-traditional trademarks such as sound, color, and scent are not protected until they acquire secondary meaning.

Suggestive marks are next on the scale of protection and require some imagination, thought, or perception to come to a conclusion as to the exact nature of the goods. One example is *SNO-RAKE* for a type of snow removal tool. Marks of this nature may be similar to a descriptive mark, but are registerable without a showing of acquired distinctiveness.

The most distinctive marks on the scale are marks that are either entirely coined, such as *EXXON*, and cannot be found in any dictionary or are arbitrary in the sense that they use common words in a way that is not expected, such as with *CAMEL* for cigarettes.

On the business side, there is a strong temptation for a brand manager to select a trademark that is descriptive for the simple reason that a descriptive mark immediately provides the customer with information about the product. However, the downside to a descriptive mark is that the terms are normally used on a wide variety of goods. If the company is able to obtain a descriptive trademark, it will be for a narrow set of goods. Descriptive marks can be problematic in the event that the product takes off in popularity and the trademark owner would like to parlay that success by expanding into other product areas.

Example: If Tastewell decides to select Tropical Island as the trademark for its new mango flavored cheese, it may be able to acquire trademark rights in that mark after using it for a period of time, provided that it was the first to adopt it for fruit flavored cheese. While this might prevent a competitor from using TROPICAL ISLAND as a mark on directly competing product, Tastewell would have difficulty preventing a third party from using Tropical Island for "crackers" or other non-cheese food products. Likewise, it may not be able to prevent a competitor from using the term "tropical" in a descriptive sense in labels or advertisements.

In contrast, an arbitrary or fanciful mark is undoubtedly a strong mark. Since it bears no relationship to the goods or services, it usually takes more time and marketing efforts to create brand awareness among consumers. Once that recognition is achieved, the company can achieve an extremely strong brand recognition. APPLE for computers and AMAZON for online retail services are two well-known examples.

3.3 Non-protectable Subject Matter

Trademarks are only protectable if they are capable of distinguishing the goods or services of one owner from those of another. Therefore, generic marks cannot be protected and descriptive marks are only protectable with a showing of acquired distinctiveness. However, there are also additional types of marks that are not protectable.

(1) *Functional marks.* Non-traditional marks such as colors and product designs are not protectable if they are, on the whole, functional. For example, the color yellow has been found to be functional for safety products.
(2) *Surname.* Marks that are primarily merely surnames are not protectable if a showing of acquired distinctiveness is absent.
(3) *Immoral or scandalous marks.* The Lanham Act specifically bars immoral or scandalous matter from protection.
(4) *Likelihood of Confusion.* Proposed use or registration of a trademark can be blocked by a prior trademark holder under common law or federal registration rights if the proposed use is identical or is likely to cause confusion with the existing trademark rights.
(5) *Geographic descriptiveness.* Marks that are geographically descriptive cannot be registered until they acquire secondary meaning. An example of such a mark is CALIFORNIA PIZZA KITCHEN.
(6) *Names and portraits.* Trademark protection is not available for marks that consist of the name, portrait, or signature of a living person without their consent.

(7) *Dilution*. A trademark that dilutes the distinctive quality of a famous mark is not registerable even if no likelihood of confusion is present.

3.4 Selecting a Trademark

In most cases, the process for selecting a trademark is not simple. Whether a company is just a start up and is looking for a new house mark or a multi-national organization looking for a mark for a new niche breakfast cereal, a process for the selection of a trademark must be in place. The basic tenants of a typical selection process are discussed below.

3.4.1 Brainstorming Phase

During this phase, creative, marketing, and technical people develop a list of potential marks. If a company does not have the time or personnel to engage in a brainstorming session, it may hire an outside company that specializes in developing potential marks. Although such companies are normally good at what they do, they can be very expensive.

3.4.2 Narrowing Phase

During this phase, either a person or committee narrows the list of marks and those on the committee should consider some of the points already discussed, including:

- The commercial appeal of the proposed mark.
- The legal strength of the proposed marks. As noted, arbitrary or suggestive marks are typically more easily registered and defended.
- Whether to add a design feature or stylization.
- Whether competitors have similar marks.

Although the marks should be ranked in order of preference, each of the marks on the final list should be acceptable to the company as a trademark.

3.4.3 Knockout Phase

Before going through the expense of a trademark search (see next step), it is advisable to perform a knockout search to identify any obvious conflicts. A knockout search may be performed in-house or by an outside attorney on the available databases of registered marks to eliminate marks that may be difficult to register or conflict with a third party's marks.

3.4.4 Clearance Search

If a mark passes the knockout phase, a clearance search is conducted in the countries or jurisdictions of interest for the marks of interest. These searches are typically conducted or overseen by either outside law firms or in-house personnel with trademark expertise. A search should include federal registrations, state registrations, common law marks, Internet domains, and websites.

When considering whether to conduct a full comprehensive search, consider the following points:

- How is the mark planned to be used?
- How widely will the mark be used?
- Will the mark be used on tooling for the product or just on advertisements?
- Will the mark be used on television?
- How important is the mark for the company?

In setting a budget for these clearance searches, costs are directly proportional to the number of countries for which protection is sought. In most cases, the difficulty of clearing a mark increase as the pool of potential problem marks increases. Therefore, a decision should be made early on as to the intended countries that will use the mark. Although using an in-house search may be cost effective, it is generally recommended to have the clearance search performed outside of the company.

3.4.5 Obtaining an Opinion

The final stage in the selection process is whether to obtain an opinion as to the availability of a mark for use and registration in conjunction with a desired mark. Although obtaining an opinion based on the results of a search is not required, there are compelling reasons to obtain such an opinion.

First, it can be hard to assess search results. This is especially true in the an industry where there are many competing products in other industries or the mark may be considered descriptive (such as the food industry).

Second, although the clearance process can be expensive, the failure to clear a mark can be even more expensive. The ramifications of selecting a potentially confusing mark are serious. An infringer faces both the prospect of monetary damages and attorney fees if found to infringe; the resulting embarrassment and loss of marketing momentum, and the finding of infringement would likely result in an injunction prohibiting use of the mark.

Third, obtaining a favorable opinion is strong evidence that there was no bad faith in using the mark and also shows that the brand manager is exercising due diligence. Both these factors are important to a court when assessing willfulness and to others who may decide to second guess a trademark's selection.

3.5 Protecting the Mark

3.5.1 Common Law of Trademark

In the United States, unlike most countries, unregistered trademarks and names enjoy common law protection. This means that the party who adopts and uses a mark in a particular geographic territory is entitled to protection against a subsequent user who adopts the same or similar mark in that same territory. The concept of "territory" is a relatively nebulous and narrow concept that depends on the nature and extent of the use of the mark in a territory. For example, a business such as a restaurant that has limited advertisement or recognition in an area may only acquire trademark rights within a limited radius of its location under common law of trademarks. In contrast, a large company that advertises nationally and has sales throughout the United States may conceivably claim trademark rights throughout the entire United States.

Although reliance on common law rights may offer initial costs savings, common law rights have several limitations as set forth below:

- limited to the particular territory where the mark was used;
- an innocent user who obtains a federal trademark registration may take over the rest of the country;
- establishing common law rights is extremely fact sensitive. Accordingly, such rights can be difficult and extremely expensive to prove in court.

Example: After receiving the trademark search results, Tastewell decides to delay filing a federal trademark application, but instead immediately begins test marketing sales of Tropical Island Cheese product in Peoria, Illinois. After 6 months of better-than-expected sales and customer reviews, Tastewell decides to launch the product nationally and seeks federal trademark protection. Unbeknown to Tastewell, however, Dairylander, a small California dairy company innocently launches a cheese product under the same trademark. As part of its marketing efforts, Dairylander advertises in all the major newspapers in California and on local television and radio as well. Due to customer demand, all major supermarkets carry Dairylander's Tropical Islands Cheese. Under this scenario, even though Tastewell is the senior party with first use, it is likely that Dairylander will be able to continue to use its market in California and perhaps in areas of surrounding states.

3.5.2 Federal Trademark Protection

Although a trademark owner can simply acquire geographic trademark rights through use of the mark, there are a number of advantages in filing for a federal trademark registration:

- *Right to use the ® symbol with all federally registered marks.* This symbol can have potent deterrent effects.
- *Provides constructive notice to the public of the claim to ownership of the mark.* Makes it much more difficult for a party to plead innocent infringement. Also, a basis for the United States Patent and Trademark Office (USPTO) to reject confusingly similar marks. Common law marks cannot be cited by the USPTO to deny registration.
- *Confers nationwide priority of rights effective from the US application filing date.* This may be the most important advantage. With this right, unlike with common law, a trademark owner does not have to prove use in a particular state or states(s) in order to claim trademark rights.
- *The legal presumption of the registrant's ownership of the mark, its validity, and the registrant's exclusive right to use the mark nationwide.* With this right the trademark owner once again does not have to prove rights in the mark, they are presumed.
- *Ability to bring an action concerning the mark in federal court and possible recovery of treble damages and attorney's fees*
- *US registration may serve as a basis to obtain registration in foreign countries without first using the mark.*
- *Ability to file the US registration with the US Customs Service to prevent importation of infringing foreign goods*
- *Availability of incontestability status after 5 years of continuous use and registration.*

Although a trademark owner may gain some rights through the use of a mark, it is recommended to file a trademark application as soon as possible and early in the development of the product.

3.5.2.1 The Federal Trademark Application Requirements

The requirements for filing a trademark application are relatively straightforward. In order for an application to receive a filing date, it must include: (1) the required filing fee for at least one class of goods or services; (2) the name of the applicant; (3) the name and address for the applicant or attorney for communication; (4) a clear drawing of mark to be registered; and (5) the identification of the goods and/or services that the mark will be used with.

The most complex part of filing an application is preparing the identification of goods and/or services for which trademark protection is being sought. The goal is to draft identification of goods and services as broadly as possible because the identification cannot be later expanded. The USPTO requires that the identification be specific and definite. Moreover, if use is claimed, it is important that the mark be used on all of the goods or services, and that there be a bona fide intent to use the mark if the application is filed on an intent-to-use basis. If these requirements are not met, any subsequent registration could be subject to cancellation for fraud on the USPTO.

There are four bases on which to register a mark:

1. Actual use in commerce;
2. Bona fide intention to use mark in commerce;
3. Foreign registration – this is only available to companies domiciled outside the United States; and
4. Under the Madrid Protocol – this is only available to companies domiciled outside the United States.[1]

In order for an applicant to claim use in commerce as basis for an application, the mark.must have been in use as of the application date. The date of first use is the date on which the goods were first sold or transported in interstate commerce or the services first rendered anywhere in the world in an arm's length transaction. The date of first use in commerce is the date that the goods are either sold or transported in commerce such that they could be regulated by applicable laws. For services, the date of first use in

---

[1] Foreign trademark protection is discussed below in this chapter.

commerce is the date the mark is first used or displayed in sales or advertising of services and the services are rendered in interstate commerce.

Once the trademark owner decides to adopt and use a trademark, an intent-to-use application should be filed as soon as possible in order to gain the advantage of constructive use date.

3.5.2.2 Examination

Following the filing of a trademark application, the USPTO assigns it to a trademark examining attorney who examines the mark to determine whether it is entitled to registration. The trademark examining attorney conducts a trademark search to determine if the mark is likely to cause confusion with any other mark on the Principal Register and reviews the application for compliance with the Trademark Act and USPTO Rules.

If for any reason the trademark examining attorney determines that the mark is not registerable for any reason, the trademark examining attorney will issue an Office Action that advises the applicant of all grounds of refusal and all matters that require further action. The applicant has 6 months to respond to the Office Action. This 6-month period runs from the mailing date of the Office Action. Failure to fully respond to the Office Action within the statutory period results in the application becoming abandoned.

Once the applicant has had the opportunity to respond to all issues raised in the Office Action, the examining attorney issues a final communication that allows or finally rejects the application. After a negative final office action, the applicant may file a request for reconsideration and submit additional evidence and argument in order to persuade the examining attorney to withdraw the final office action. If the examiner fails to withdraw the final office action within 6 months, the applicant must either meet every requirement of the office action or appeal to the Trademark Trial and Appeal Board in order to avoid abandonment of the application.

Following examination, if it appears that a mark is entitled to registration on the Principal Register and there are no outstanding requirements or refusals, the examining attorney will approve the mark for publication in the USPTO *Official Gazette*. If a third party does not file an opposition within 30 days of publication or request a time extension to file an opposition, the application will proceed to registration.

For applications based on intention to use and where no Amendment to Allege Use has been filed before publication, the USPTO will issue a Notice of Allowance. The application will proceed to registration upon the filing of a Statement of Use. A Statement of Use or request for extension of time to file a Statement of use must be filed within 6 months of the mailing date of the Notice of Allowance. The applicant may request an extension for a

6-month period without showing good cause. Thereafter, the applicant may receive an additional 6-month extension upon request and by a showing of good cause.

3.5.3 State Registrations

State laws also provide for trademark registrations. However, given the many advantages to Federal Registration, there is little point to obtaining a state registration unless the trademark owner cannot establish use in interstate commerce or has a specific legal need to take advantage of that state's trademark or anti-dilution remedies.

3.5.4 Maintaining Rights

Trademark rights can be a company's most valuable asset. Like any asset, these rights need to be protected and maintained. Trademark rights are lost when the mark no longer acts as an identifier of the goods or services. This can occur through abandonment from non-use or through a course of conduct including acts of commission or omission which allow a mark to become the generic name of goods or services or lose significance as a mark.

3.5.4.1 Maintaining Federal Registrations

Section 8 of the Lanham Act requires that an affidavit or declaration verifying continued use in commerce or excusable non-use due to special circumstances be filed with the USPTO between the fifth and sixth anniversary and on or between the ninth and tenth anniversary date of the registration. This requirement applies to all registrations. Failure to meet this requirement will result in cancellation of a registration.

The duration of a trademark registration has varied over the years. Since November 16, 1989, however, all registrations that have been issued or renewed after that date have only a 10-year term. Therefore, in order to maintain a federal trademark registration, a renewal application must be filed on or between the ninth and tenth anniversary date of the registration.

3.5.4.2 Licensing

A trademark owner can license its trademark to a third party. However, a trademark owner must reserve the power to exercise quality control over the nature and quantity of the goods and services in a license. If the trademark owner fails to exercise such quality control, the license is considered a "naked license" and the mark may be abandoned.

3.5.4.3 Assignments

A trademark may only be assigned to another party with the goodwill of the business in which the mark is used. Failure to meet this requirement can result in the trademark becoming void. Similarly, intent-to-use applications cannot be assigned prior to the filing of a Statement of Use or Amendment to Allege use unless the assignment is to the successor to the ongoing and existing business of the applicant or to the portion to which the mark pertains.

3.5.4.4 Genericide

There are many examples where once valuable trademarks have become generic, i.e., the trademarks have ceased to function as a source indicator and have become the name of a particular type of a product. The following terms were at one time a company's trademark:

- ALE HOUSE (for restaurant and bar services)
- ASPIRIN (for acetyl salicylic acid pain reliever)
- CELLOPHANE (for transparent cellulose sheets and films)
- COLA (for soft drink)
- CRAB HOUSE (for seafood restaurant)
- CUBE STEAK (for steaks)
- DERBY PIE (for chocolate nut pie)
- ESCALATOR (for moving stairs)
- FONTINA (for cheese)
- HOAGIE (for a sandwich)
- HONEY BROWN (for a brown ale made with honey)
- JUJUBES (for gum candy)
- LIGHT BEER (for beer with fewer calories)
- MONTESSORI (for educational services)
- MURPHY BED (for folding bed)
- SHREDDED WHEAT (for baked wheat cereal)
- SOFTCHEWS (for chewable medical pills)
- SUPER GLUE (for glue)
- SURGICENTER (for surgical center)
- TEDDY (for bear toy)
- THERMOS (for vacuum-insulated bottles)
- TRAMPOLINE (for jumping and gymnastic equipment)
- YELLOW PAGES (for business telephone directory)
- YO-YO (for toys)

These terms are now generic and available for use by all. If a trademark owner wishes to avoid that result, it is essential that all those associated with a trademark understand the requirements. The following tips can help protect the value of a trademark.

- *Do not trademark the name of your product.* If you cannot provide the generic name of your product without referring to your brand, then the trademark may become generic.
- *When using a trademark in a sentence, always use the trademark as an adjective.* If this rule is not followed, the public may come to see the trademark as the generic name of the product or service which is what happened to former trademarks such as escalator, cellophane, and kerosene. Use of the word "brand" also can help emphasize that a term is a mark and not a generic descriptor, for example: Jell-O® brand gelatin dessert. There is one caveat to this rule – many companies use their trade names as trademarks. In those instances where the trade name is being used, it is a proper noun, not an adjective.
- *In a sentence, the trademark should be set apart from the text in some fashion.* This can be accomplished by using ALL CAPITAL LETTERS, **bold face type**, Initial Capital Letter, *italics*, or through the use of a unique font.
- *Monitor use of your mark.* If you see your mark starting to appear in all lower cases in publications, this is a danger sign. Steps should be taken to send notifications to advertisers using the mark in that way.
- *Police misuse.* If your trademark is found in a dictionary, whether intentional or unintentional, it is strong evidence that the trademark is generic. Corrective action in the form of a letter to the publisher should be taken immediately. Likewise, if you see your mark appear in lower case letters this is a danger sign.
- *Use the brand in a consistent manner.* Not only is recognition of the mark enhanced through consistent use, inconsistent use may confuse consumers, dilute the distinctiveness of the mark, and lead to abandonment of the mark.
- *Use the appropriate trademark designation.* In the United States the symbol TM can be used to identify an unregistered trademark, and SM can be used to identify an unregistered service mark, and ® can be used to identify a registered trademark or service mark. Many foreign countries use similar terms. Local laws should be consulted because many countries require proper use of the symbol in their country in order to collect damages for infringement and such symbols may not be the same as those accepted in the United States.
- *Develop a trademark usage manual.* All companies should have a manual to advise employees and others on the proper use of the company's trademarks.

- *Non-use.* A trademark owner must use the mark to maintain it. Three years of non-use results in a presumption of abandonment.
- *Infringement.* The key function of a trademark is as a source identifier. If the same or similar trademark is used by more than one company on the same or related goods, the mark may cease to be a source identifier. Therefore, it is important for a company to police third party usage of its marks and take appropriate action ranging from cease and desist letters to legal action.

3.5.5 International Protection

Trademark protection is also available internationally. Unlike the United States, most countries award trademark rights solely on a first-to-file basis. Therefore, it can be extremely important to consider the need to file a trademark application in other countries. A US registration can form the basis of an application in a foreign country without the necessity of first having used the mark in that country. In most countries, a US company can file a trademark application that claims the same US application date if the trademark application is filed within 6 months of the US filing date. However, foreign trademark protection is often very expensive and the costs multiply depending on the jurisdictions in which the protection is sought. Accordingly, a company should develop a list of countries for which registration will be sought.

When considering which countries to pursue protection, the following factors should be evaluated.

- Serious consideration should be given to registering in those countries where the mark is used or will be used in the near term.
- Consideration should also be given to countries where the company is planning on expanding in the next 3–5 years. This is especially true for large markets such as China which uses a first-to-file system.
- Finally, if counterfeiting is a problem, consideration should be given to defensive filings in some of the key counterfeiting source nations such as Taiwan, China, and Vietnam.

Once the decision is made, a trademark owner has several options for filing overseas. In many cases, the filing is done in each individual country which requires hiring a trademark attorney in each country and paying filing fees in each country. However, there are some international treaties that allow a trademark owner to avoid some of these fees and costs.

3.5.5.1 Madrid Protocol

Foreign filing costs can be reduced using the Madrid Protocol. The Madrid Protocol is an international trademark treaty which permits the owner of a "home country" registration to file an international application with its national trademark office that designates other member countries. The Madrid Protocol offers cost savings and increased efficiency for US trademark holders. The International Trademark Association has aptly summarized the benefits of the Madrid Protocol as offering:[2]

- one application;
- in one place;
- with one set of documents;
- in one language;
- with one fee;
- resulting in one registration;
- with one number;
- one renewal date;and
- covering more than one country.

The cost savings of registration through the Madrid Protocol are significant. Another advantage of the Madrid Protocol is the simplicity in filing application amendments. Without the Madrid Protocol, applications would have to be filed and prosecuted individually in every country in which the mark is registered. However, the Madrid Protocol simplifies this process at a reduced cost.

Other advantages of an international registration under the Madrid Protocol include having priority of protection in all designated countries from the date of international registration, as opposed to the date of registration in the individual countries. Also, the Madrid Protocol limits the time a national office has to act once it receives a request for extension of a Madrid registration. If the office does not act to oppose protection during the allotted time, the registration is automatically granted.

By offering simultaneous registrations in the United States and foreign countries, the Madrid Protocol also reduces trademark piracy. Without the Madrid Protocol, individuals in the foreign countries are often free to register a US company's trademark, and attempt to sell the mark to the US company at highly inflated prices. Registration under the Madrid Protocol

---

[2]International Trademark Association, *The Madrid Protocol: Impact of U.S. Adherence on Trademark Law and Practice*, at p. 1 (Revised April 2003).

reduces such piracy since all designated countries are given the same priority date.

Notwithstanding the cost savings and increased efficiency associated with the Madrid Protocol, there are some drawbacks for US applicants. First, the Madrid Protocol requires that the scope of goods or services covered by the registration be limited to the home country's registration rules. US applicants that seek registration through the Madrid Protocol will be prejudiced in this respect since the USPTO requires more detailed identification of goods and services than most other countries. Unlike many other countries, the USPTO will not accept registration of marks for broad classes of goods and services. Therefore, US companies may limit the scope of protection that could otherwise be obtained in other countries by filing an international registration as opposed to filing individual national applications.

Another potential consequence of the Madrid Protocol for US trademark owners is the limitation the USPTO imposes on applicants to provide a statement of use or *bona fide* intent to use the mark in commerce before obtaining a filing date, and proof of use in commerce before a registration will be issued. Most other countries do not require a similar statement of use or intent to use the mark in commerce, or proof of such use. Therefore, US trademark applicants may be disadvantaged by their inability to "reserve" a trademark under USPTO procedure.

Trademark owners filing under the Madrid Protocol are also subject to "central attack." If a home country application or registration is cancelled or abandoned during the first 5 years of registration, whether completely or partially, the home country must notify WIPO. The international registration then lapses with respect to all designated countries. This is particularly disadvantageous to US trademark owners because there are usually more grounds for challenging registrations under US law than in other countries.

The Madrid Protocol provides for a partial safeguard against "central attack" by providing a 3-month grace period for the owner of a cancelled registration to file national applications in designated countries that enjoy the same priority as the international registration. This process, however, can be expensive and time consuming.

Unlike national application systems, under the Madrid Protocol an assignment may only be recorded if the assignee is itself qualified to file a Madrid Protocol application. Although this only affects the recording of the assignment and national laws will govern the legal effect of the assignment, member countries such as the United States have passed laws which make assignments to non-member citizens invalid. This assignment provision may be problematic in cases where a US citizen or corporation wishes to assign registration(s) to a non-member citizen or corporation for tax or other purposes.

The final drawback of implementation of the Madrid Protocol for US trademark owners is that the system is outside of the United States and the procedures can seem unfair for those used to US filings. For example, the time period for responding to office actions under the Madrid Protocol may be quite short due to the fact that an office action is sent from the national office to WIPO and WIPO sends it to the trademark owner. Moreover, many Madrid Protocol countries do not send a Registration Certificate or other notice once the registration is issued. Therefore, the trademark owner is left in the uncertain position of not knowing for sure if the registration has actually issued.

3.5.5.2 European Community Trademark

The European Community system offers a trademark system that allows for registration of a trademark in all of the member countries for one application filing fee.[3] If the trademark owner intends to use the mark in more than two European Community member countries, it is typically more cost effective to file for a Community Registration. Although a Community Registration can be canceled for 5 years of non-use, use in one country is enough to satisfy the use requirement.

---

[3]Member countries include: Austria, Benelux (Belgium, the Netherlands and Luxembourg), Bulgaria, Cyprus, the Czech Republic, Denmark, Estonia, Finland, France, Germany, Greece, Hungary, Ireland, Italy, Latvia, Lithuania, Malta, Poland, Portugal, Romania, the Slovak Republic, Slovenia, Spain, Sweden and the United Kingdom.

Chapter 4
Copyrights

4.1 Copyrightable Subject Matter and Scope of Protection

The federal Copyright Act protects authors' creative works. "Copyright protection subsists...in original works of authorship...fixed in a tangible medium of expression ...."[1] Copyright automatically arises upon creation and fixation of an original work, which in the food technology industry may include food advertising and marketing materials/packaging, or secret and other materials including recipes and software.

There are limitations on the scope of copyright protection. First, while registration is not required to secure copyright protection, it is a prerequisite to litigating a copyright claim and is desirable to preserve the remedies for attorney's fees and statutory damages. Second, "[i]n no case does copyright protection for an original work of authorship extend to any idea, procedure, process, system, method of operation, concept, principle, or discovery...."[2] Third, where a "useful article" is concerned, copyright protection is only afforded to the extent the original creative expression is physically or conceptually separable from the utility of the useful article. Finally, certain subject matters are not subject to copyright.

4.1.1 Foods as Copyrightable Subject Matter

An original sculpture is copyrightable whether it is made of granite, bronze, chocolate, butter, or ice. In the case of the edible media, however, there can be a "fixation" issue. If an ice sculpture is created on a hot day knowing the sculpture will melt, it is temporal and not fixed. Thus, its creator might use

---

[1] 17 U.S.C. § 102(a).
[2] 17 U.S.C. § 102(b).

R.W. O'Donnell et al., *Intellectual Property in the Food Technology Industry*,
DOI: 10.1007/978-0-387-77389-6_4, © Springer Science+Business Media, LLC 2008

photography or make a mold to satisfy the Copyright Act's fixation require-ment. This is true because even if the original is destroyed, the intangible copyright endures to the extent it has been preserved in some form from which the work can be reproduced.

For mass produced foods, generally a distinctive artistic shape will be reproduced using a mold. The availability of copyright protection depends on whether there is sufficient original expression in the overall appearance of the item. For example, the Copyright Office might deny protection for a simple triangular-shaped ice-cream bar, but grant protection for an original combination of five different geometric shapes, each of a different color and texture.

In a restaurant setting, signature dishes that embody a great deal of orig-inal artistic presentation are copyrightable to the extent a fixation is made, such as by taking pictures or making plastic replicas. The pictures or replicas can then be used for marketing and advertising purposes and may acquire distinctiveness in a trademark sense as well.

4.1.2 Advertising, Marketing Materials, and Packaging

Original authorship is often embodied in advertising and marketing materials and packaging. The primary issues of concern in these areas are copyright ownership, avoiding infringement, and preserving remedies through regis-tration.

Through registration, a copyright owner preserves the ability to recover attorney's fees and statutory damages, remedies that are not available if infringement commences before registration. The registration process also compels an examination of authorship/ownership issues as well as necessi-tating a review of whether any preexisting material is embodied in a work which may give rise to a need for obtaining permission for use of such material.

Example: Tastewell is launching a new cheese product. The product is a cheese with papaya and mango slices. The cheese will be sold in the shape of a papaya, and packaged in plastic packaging. Tastewell's mar-keting department, in connection with the advertising firm Independent Contractor's Inc., has created packaging and labeling having a tropical theme (palm trees, beaches, sun, etc.) that includes "Papa Mango," a character whose body is a mango, and whose arms are papaya slices.

In the example, it would be wise to register copyright in the original character Papa Mango. For example, the California Raisin Advisory Board had an advertising campaign featuring "*The California Raisins*," that spawned a business of plush figures, T-shirts, performances and other items which in 1988 exceeded the sales of raisins by the California farmers.[3]

Following in the tradition of *The California Raisins*, Tastewell may decide to pursue a TV commercial featuring Papa Mango doing the "*Macarena*." Pursuing registration of the copyright in the commercial should raise the issue of identifying all of the "authors" of the commercial and the use of preexisting material, such as the song *Macarena* by Los del Río. This will normally bring to light the need for appropriate assignments and clearances.

To the extent that packaging embodies creative expression, packaging is protectable under copyright. This can extend to the shape of packaging such as a Mickey Mouse-shaped popsicle. Containers are, however, generally useful articles that would be more appropriately protected by patents. In some instances, distinct containers may also be protected under trademark registrations.

Even if the Papa Mango character is not directly copied by a competitor, copyright registration would facilitate addressing wrongful use of a derivative work.

Example: Tastewell's competitor, Bland Foods Corp., may decide to launch a competing cheese with pieces of passion fruit in the cheese. If Bland Foods also uses a tropical theme along with a "Mama Papaya" character whose body is a papaya, and whose arms are mango slices, Tastewell may have a copyright claim based on its rights to derivative works.

4.1.3 Secret and Other Materials Including Recipes and Software

In many instances, a business may develop recipes or software related to food products. To the extent such works contain original expression, copyright law protects them with the caveat that the underlying methods and procedures are not protected by copyright. Even software with methods therein may be protectable and enforceable, provided that duplication can be proven.

---

[3] *See Wikipedia, the free encyclopedia*, for additional details.

For software, owners often want to ensure secrecy and copyright allows such secrecy. When an owner needs secrecy of works such as recipes and software source code, it can register the work by providing sufficient identifying material instead of the normally required copy of the work. The identifying material may have confidential portions redacted to preserve secrecy.

Example: Two of Tastewell's senior scientists, Dr. Curd and Dr. Whey, conceive the idea for the new cheese product with fruit. Several younger scientists in the R&D group are involved in the analysis and testing of the product, and one of the younger scientists discovered that the product has a longer shelf life by increasing the heat exposure during processing. Dr. Curd resigned from Tastewell shortly after completing the initial conception of the cheese product, and began working for Tastewell's competitor, Bland Foods.

In advance of the departure of Dr. Curd, Tastewell could proceed with the filing of a copyright application with respect to the development materials to officially document Tastewell's development efforts and Dr. Curd's contributions to the new product developments. Redacted versions of the material can be submitted for the registration process to maintain confidentiality. However, sufficient material should be provided to serve as evidence of Dr. Curd's contributions to and knowledge of Tastewell's developments should the need arise.

4.2 Ownership/Authorship

Three simple rules control the initial ownership of copyright. First, in general, ownership of copyright "vests initially in the author or authors of the work." [4] Thus, a person who creates an original work automatically becomes its copyright owner. Second, and a major exception to the first rule, where a work is made for hire, the employer or commissioning party is deemed to be the author. Third, where there is more than one author, "the authors of a joint work are co-owner of copyright in the work." [5]

Although these rules are relatively simple, determining authorship can be difficult. When authors cannot agree on authorship, the Copyright Office will register conflicting claims to copyright in the same work. To avoid such

---

[4] 17 USC § 201(a).
[5] 17 USC § 201(a).

a dispute, it is always best to try to resolve authorship/ownership issues at an early stage. For an employer, employment contracts should be used to define scope of employment and any work being done by independent contractors should be covered by an agreement that assigns any copyrights.

> *Example*: With respect to Tastewell's new fruit and cheese product noted above, the respective efforts of the marketing department of Tastewell, the advertising firm, Independent Contractor's Inc., in connection with the tropical theme packaging, and labeling and creation of the "Papa Mango" character need to be evaluated, to avoid problems with ownership of the copyright in such materials.

4.2.1 Works for Hire

The Copyright Act in 17 USC § 101 defines:
A "work made for hire" is–

(1) a work prepared by an employee within the scope of his or her employment; or
(2) a work specially ordered or commissioned for use as . . . [specific list of types of items]. . .if the parties expressly agree in a written instrument signed by them that the work shall be considered a work made for hire.

Employers need not have an agreement with their employers to create a "work made for hire" obligation. Employment contracts, however, are often useful in settling questions involving whether a particular work was prepared in the "scope of employment." Clear policies often avoid problems in this respect. On the other hand, the law requires specially ordered and commissioned works to be identified in writing as "work for hire." In addition, even if there is a written agreement as to "work for hire" status, only the specifically enumeratedtypes of works are eligible to be considered "work for hire."

4.2.2 Jointly Authored Works

17 USC § 101 of the Copyright Act defines a "joint work" as "a work prepared by two or more authors with the intention that their contributions be merged into inseparable or interdependent parts of a unitary whole." The authors of a joint work may be natural persons, persons, or other entities by

virtue of "works for hire" or a combination of both where the some contributions are "work for hire" and other contributions are not "works for hire." In the case of multiple employees creating a work for the same employer, there is only one author, the employer, so that such a work is not a joint work.

For works that are jointly owned, unless there is an agreement to the contrary, any joint owner is free to exploit the copyright in the entire work. The other joint owners, however, have a right of contribution to the profits made from the exploitation of a work.

4.2.3 Copyright Transfers: Assignment and Licensing

After initial ownership of copyright is established through authorship, copyrights may be transferred. The law requires a writing, however, to assign a copyright. 17 USC § 204(a) provides:

> A transfer of copyright ownership, other than by operation of law, is not valid unless an instrument of conveyance, or a note or memorandum of the transfer, is in writing and signed by the owner of the rights conveyed or such owner's duly authorized agent.

Proper recordation of copyright assignments with the Copyright Office should be made to perfect title to copyrights. The requirement for a writing under 17 USC § 204 does not extend to copyright licenses. Accordingly, there can be oral and implied copyright licenses. Recordation of license rights is not required.

In the case of works not made for hire, both assignments and licenses can be terminated after 35 years. [6]

Example Continued: Referring to the above example, generally all of the materials developed by Tastewell's marketing department employees are "works for hire" that are deemed authored and owned by Tastewell upon their creation. The copyright in Independent Contractor Inc.'s contributions to the marketing and advertising materials, however, will only be owned by Tastewell if there is an assignment of rights. The situation becomes more complex if Independent Contractor Inc. hires others to work on the Tastewell account on an independent contractor basis. For example, Independent Contractor Inc. might engage

---

[6] 17 USC § 203.

someone like Michael Brunsfeld, the artist responsible for the original design of the California Raisins, to bring to life Papa Mango.

In view of the rules governing copyright ownership, Tastewell would normally wish to specify in a contract engaging Independent Contractor Inc. that all IP rights in works created in the course of the services provided to Tastewell are and will be assigned to Tastewell. The provisions should compel Independent Contractor Inc. to make sure that persons or entities which Independent Contractor Inc. engages to work on Tastewell's projects likewise agree to assign copyrights and all other IP rights. Typically, such an agreement would also obligate the execution of future assignments and other documents needed to perfect title in any associate IP rights.

If "Papa Mango" was created by the joint efforts of Independent Contractor Inc.'s employees and an independent contractor, Mr. Artist, Tastewell would need assignments from both Independent Contractor Inc. and Mr. Artist to obtain complete copyright ownership of "Papa Mango."

4.3 Derivative Works

One of the informational requirements in registering copyright in a work is to identify preexisting copyrighted material that is present in the copyright application. This is important for registration purposes because copyright in a derivative work is limited to the newly added material; the preexisting material is protected by its own prior copyright.

To the extent a work "unlawfully" uses preexisting material, the law may invalidate copyright protection.

4.4 Fair Use

The Copyright Act permits certain copying under the doctrine of fair use. The copyright Act defines a "fair use" balancing test in 17 USC § 109 that provides:

> . . .the fair use of a copyrighted work, including such use by reproduction in copies or phonorecords or by any other means specified by that section, for purposes such as criticism, comment, news reporting, teaching (including multiple copies for classroom use), scholarship, or research, is not an infringement of copyright.

In determining whether the use made of a work in any particular case is a fair use the factors to be considered shall include –

(1) the purpose and character of the use, including whether such use is of a commercial nature or is for non-profit educational purposes;
(2) the nature of the copyrighted work;
(3) the amount and substantiality of the portion used in relation to the copyrighted work as a whole; and
(4) the effect of the use upon the potential market for or value of the copyrighted work.

The fact that a work is unpublished shall not itself bar a finding of fair use if such a finding is made upon consideration of all the above factors.

What does and does not constitute "fair use" is often subject to debate. A good rule of thumb is that if you would object to someone copying what you are considering, you may not want to copy without talking to your attorney.

Example: Where Tastewell's competitor, Bland Foods Corp., launches a competing cheese with pieces of passion fruit in the cheese using a tropical theme along with a "Mama Papaya" character whose body is a papaya, and whose arms are passion fruit slices, Bland Foods may try to rely of a "fair use" defense. The commercial nature of the use, however, would weigh against an assertion that the "Mama Papaya" character whose body is a papaya, and whose arms are passion fruit slices as not an infringing derivative work of the "Papa Mango" character whose body is a mango, and whose arms are papaya slices.

Where significant use of a third parties work is desired to be made, it is generally advisable to obtain permission or a license. For music, the copyright act provides for compulsory licensing. For major projects, clearance activity may entail contacting multiple parties from collective rights groups such as the Copyright Clearance Center, ASCAP, and BMI to track down individual authors.

Example: As noted above, the copyright in Independent Contractor Inc.'s contributions to the marketing and advertising materials, however, will only be owned by Tastewell if there is an assignment of rights. Warranties as to originality and rights to use preexisting materials are normally desired for clearance purposes. For example, Tastewell would specify that Independent Contractor Inc. warrant that material it provides is clear for Tastewell's desired usage. Tastewell may require Independent Contractor Inc. to identify all preexisting materials and to verify that permissions were obtained or to provide an explanation why the use of such preexisting material is "fair use." Tastewell may also require Independent Contractor Inc. to identify all contributions made by persons or entities that are not employees of Independent Contractor Inc.

4.5 Registration Issues

Copyright registrations provide an invaluable source of information about copyrighted works and aid in the preservations of works and enhancement of the Library of Congress's collection. Moreover, for copyright owners, the registration process provides a valuable procedure to compel an examination of the issues of copyright ownership and the use of third party materials.

Copyright registration is a deceptively simple procedure that leads to questions of authorship and ownership.

Registration of copyright preserves important remedies, including statutory (mandatory) damages and attorney fees; in fact, a plaintiff cannot even bring a copyright action without a copyright registration.

Example: The following example Copyright Application for registration of the copyright in the "Papa Mango" character is based upon the creation of "Papa Mango" through the joint efforts of Independent Contractor Inc.'s, employees and an independent contractor, Mr. Artist. It reflects an assignment to Tastewell from both Independent Contractor Inc. and Mr. Artist. The assignment document should be recorded with the Copyright Office to complete the official record.

FEE CHANGES

Fees are effective through June 30, 2002. After that date, check the Copyright Office Website at www.loc.gov/copyright or call (202) 707-3000 for current fee information.

FORM VA
For a Work of the Visual Arts
UNITED STATES COPYRIGHT OFFICE

REGISTRATION NUMBER

VA   VAU

EFFECTIVE DATE OF REGISTRATION

Month   Day   Year

DO NOT WRITE ABOVE THIS LINE. IF YOU NEED MORE SPACE, USE A SEPARATE CONTINUATION SHEET.

1   TITLE OF THIS WORK ▼   NATURE OF THIS WORK ▼ See instructions

PREVIOUS OR ALTERNATIVE TITLES ▼

Publication as a Contribution If this work was published as a contribution to a periodical, serial, or collection, give information about the collective work in which the contribution appeared.   Title of Collective Work ▼

If published in a periodical or serial give: Volume ▼   Number ▼   Issue Date ▼   On Pages ▼

2   **a**   NAME OF AUTHOR ▼   DATES OF BIRTH AND DEATH
Year Born ▼   Year Died ▼

NOTE

Under the law, the "author" of a "work made for hire" is generally the employer, not the employee (see instructions). For any part of this work that was "made for hire" check "Yes" in the space provided, give the employer (or other person for whom the work was prepared) as "Author" of that part, and leave the space for dates of birth and death blank.

Was this contribution to the work a "work made for hire"?   ☑ Yes   ☑ No

Author's Nationality or Domicile
Name of Country
OR { Citizen of ▶
Domiciled in ▶

Was This Author's Contribution to the Work
Anonymous? ☑ Yes ☑ No
Pseudonymous? ☑ Yes ☑ No

If the answer to either of these questions is "Yes," see detailed instructions.

NATURE OF AUTHORSHIP Check appropriate box(es). **See instructions**

☑ 3-Dimensional sculpture   ☑ Map   ☑ Technical drawing
☑ 2-Dimensional artwork   ☑ Photograph   ☑ Text
☑ Reproduction of work of art   ☑ Jewelry design   ☑ Architectural work

b   NAME OF AUTHOR ▼   DATES OF BIRTH AND DEATH
Year Born ▼   Year Died ▼

Was this contribution to the work a "work made for hire"?   ☑ Yes   ☑ No

Author's Nationality or Domicile
Name of Country
OR { Citizen of ▶
Domiciled in ▶

Was This Author's Contribution to the Work
Anonymous? ☑ Yes ☑ No
Pseudonymous? ☑ Yes ☑ No

If the answer to either of these questions is "Yes," see detailed instructions.

NATURE OF AUTHORSHIP Check appropriate box(es). **See instructions**

☑ 3-Dimensional sculpture   ☑ Map   ☑ Technical drawing
☑ 2-Dimensional artwork   ☑ Photograph   ☑ Text
☑ Reproduction of work of art   ☑ Jewelry design   ☑ Architectural work

3   **a** Year in Which Creation of This Work Was
Completed   This information must be given ◀ Year in all cases.

b Date and Nation of First Publication of This Particular Work
Complete this information ONLY if this work has been published.   Month ▶ _____ Day ▶ _____ Year ▶ _____ ◀ Nation

4   COPYRIGHT CLAIMANT(S) Name and address must be given even if the claimant is the same as the author given in space 2. ▼

See instructions before completing this space.

Transfer If the claimant(s) named here in space 4 is (are) different from the author(s) named in space 2, give a brief statement of how the claimant(s) obtained ownership of the copyright. ▼

APPLICATION RECEIVED

ONE DEPOSIT RECEIVED

TWO DEPOSITS RECEIVED

FUNDS RECEIVED

DO NOT WRITE HERE OFFICE USE ONLY

MORE ON BACK ▶   • Complete all applicable spaces (numbers 5-9) on the reverse side of this page.
                 • See detailed instructions.   • Sign the form at line 8.

DO NOT WRITE HERE
Page 1 of _____ pages

EXAMINED BY	FORM VA
CHECKED BY	
☐ CORRESPONDENCE Yes	FOR COPYRIGHT OFFICE USE ONLY

DO NOT WRITE ABOVE THIS LINE. IF YOU NEED MORE SPACE, USE A SEPARATE CONTINUATION SHEET.

PREVIOUS REGISTRATION Has registration for this work, or for an earlier version of this work, already been made in the Copyright Office?

☐ Yes ☐ No If your answer is "Yes," why is another registration being sought? (Check appropriate box.) ▼

a. ☐ This is the first published edition of a work previously registered in unpublished form.

b. ☐ This is the first application submitted by this author as copyright claimant.

c. ☐ This is a changed version of the work, as shown by space 6 on this application.

If your answer is "Yes," give: **Previous Registration Number** ▼   **Year of Registration** ▼

5

DERIVATIVE WORK OR COMPILATION Complete both space 6a and 6b for a derivative work; complete only 6b for a compilation.

a. **Preexisting Material** Identify any preexisting work or works that this work is based on or incorporates. ▼

b. **Material Added to This Work** Give a brief, general statement of the material that has been added to this work and in which copyright is claimed. ▼

6

a See instructions before completing this space.

b

DEPOSIT ACCOUNT If the registration fee is to be charged to a Deposit Account established in the Copyright Office, give name and number of Account.

Name ▼   **Account Number** ▼

CORRESPONDENCE Give name and address to which correspondence about this application should be sent.   Name/Address/Apt/City/State/ZIP ▼

7

a

b

Area code and daytime telephone number ▶ (   )   Fax number ▶ (   )

Email ▶

CERTIFICATION* I, the undersigned, hereby certify that I am the

check only one ▶ {
☐ author
☐ other copyright claimant
☐ owner of exclusive right(s)
☐ authorized agent of _____
Name of author or other copyright claimant, or owner of exclusive right(s) ▲

of the work identified in this application and that the statements made by me in this application are correct to the best of my knowledge.

8

Typed or printed name and date ▼ If this application gives a date of publication in space 3, do not sign and submit it before that date.

_____   Date ▶ _____

Handwritten signature (X) ▼

X _____

Certificate will be mailed in window envelope to this address:	Name ▼	**YOU MUST:** • Complete all necessary spaces • Sign your application in space 8
	Number/Street/Apt ▼	**SEND ALL 3 ELEMENTS IN THE SAME PACKAGE:** 1. Application form 2. Nonrefundable filing fee in check or money order payable to *Register of Copyrights* 3. Deposit material
	City/State/ZIP ▼	**MAIL TO:** Library of Congress Copyright Office 101 Independence Avenue, S.E. Washington, D.C. 20559-6000

As of July 1, 1999, the filing fee for Form VA is $30.

9

*17 U.S.C. § 506(e): Any person who knowingly makes a false representation of a material fact in the application for copyright registration provided for by section 409, or in any written statement filed in connection with the application, shall be fined not more than $2,500.

June 1999—100,000
WEB REV: June 1999

♻ PRINTED ON RECYCLED PAPER

☆U.S. GOVERNMENT PRINTING OFFICE: 1999-454-879/71

Chapter 5
Domain Names

5.1 Securing Intellectual Property in Domain Names

Domain names are essential to promoting corporate identity and product awareness in the modern era and should be regarded like any other valuable corporate assets. A domain name is a string of unique characters used as an address to identify a particular computer or server on the Internet. For example, *fda.gov* is used to identify the United States Food and Drug Administration website, *uspto.gov* is used to identify the United States Patent and Trademark Office website, and *coca-cola.com* is used to identify the Coca-Cola Company website.

Domain names consist of a number of domain levels. For example, in a two-level domain name such as *fda.gov*, the portion of the domain name to the right of the period (i.e., "gov") is the top-level domain (TLD), and the portion of the domain name to the left of the period (i.e., "fda") is the second-level domain (SLD). Many internet users also recognize the three-letter string "www." preceding the domain name. This portion of the domain is typically considered a "subdomain" which is selected by the host computer. The second-level domain is usually selected by the user, and is typically used as a source identifier in the domain name. The second-level domain can consist of a trademark or service mark. The most common top-level domains are termed global or generic TLDs (gTLDs). The most recognized gTLDs are identified below:

- *.com* – commercial enterprises
- *.net* – networks
- *.org* – non-profit organizations
- *.biz* – businesses
- *.edu* – educational institutions
- *.gov* – US government entities

R.W. O'Donnell et al., *Intellectual Property in the Food Technology Industry*,
DOI: 10.1007/978-0-387-77389-6_5, © Springer Science+Business Media, LLC 2008

- *.mil* – US military
- *.int* – international organizations

In order to secure a domain name, it must first be registered with an ICANN accredited registrar. The Internet Corporation for Assigned Names and Numbers (ICANN) is responsible for the global coordination of the Internet's system of unique identifiers, including domain names. There are over 150 registrars accredited by ICANN which can be found at *icann.org/registrars/accredited-list.html*. Accredited registrars and domain name holders must implement and follow ICANN's Uniform Domain Name Dispute Resolution Policy ("UDRP"). Domain names are typically granted by registrars on a first come basis. Registrars are not responsible for determining whether a domain name registrant has the right to obtain a domain name. For example, if the registered domain name incorporates a trademark owned by another, the domain name registrant could be liable for trademark infringement. Therefore, it is equally important that registrants conduct a clearance search prior to adopting or using a domain name.

Example: Tastewell Industries decides to sell its new mango flavored cheese product under the name Tropical Island Cheese. In order to promote product awareness, Tastewell registers the domain name tricheese.com and begins to market the Tropical Island Cheese trademark in connection with this domain name. One year into its marketing campaign, Tastewell learns that another dairy company markets a cheese product consisting of a blend of three cheeses under the federally registered trademark TRI-CHEESE.

In this scenario, Tastewell may need to change its domain name or work out a licensing arrangement in order to avoid an infringement suit. Had Tastewell performed a trademark clearance search prior to using its tricheese.com domain, it could have chosen a different domain name and avoided the potential costs associated with having to subsequently change its domain after investing in its marketing campaign.

Registering a domain name with a registrar does not grant trademark protection for the domain. In order to seek trademark protection for a domain name, the owner must file an application with the US Patent and Trademark Office. In order to obtain a federal trademark registration for a domain name, the applicant must satisfy all of the legal requirements for registrable trademarks. In addition, the US Patent and Trademark Office has specific procedures for applications to register domain names. In registering a domain name as a trademark, the top-level domain (TLD) and subdomain (e.g., www.) are usually ignored. The USPTO typically considers only the second-level domain (SLD) when examining the mark for the likelihood of confusion with other marks. Where the domain name is used only as an Internet address, and not to identify the source of goods or services, the trademark is not registrable. However, if the SLD is an actual tradename for a product or the domain name owner uses the domain name to advertise its goods or services, the US Patent and Trademark is more likely to find the trademark registrable.

Example: Tastewell Industries has two domain names which it uses to advertise its new mango flavored cheese product sold under the name Tropical Island Cheese. The first domain name is tropicalisland-cheese.com, and the second domain name is foodforu.com. Tastewell wants to know the likelihood of obtaining a federal trademark registration for either of these websites.

In this scenario, it is likely that the domain name tropicalisland-cheese.com would more likely be considered registrable by the USPTO as long as the mark satisfied the remaining requirements for trademark applications since it uses the actual name for Tastewell's underlying product. Tastewell will likely have a more difficult time registering its foodforu.com domain name since it does not directly identify the source of goods or services.

5.2 Domain Name Disputes

Domain name disputes can arise in a number of ways. For example, use of another trademark in a domain name can be subject to an action for trademark infringement, unfair competition, or dilution of a trademark. "Cybersquatting" is another type of act that has been subject to dispute. Cybersquatting is typically considered the act of registering a domain for the purpose of preventing a trademark owner from using it in order to extract payment from the trademark owner. In addition to holding the domain name hostage to the trademark owner, cybersquatting also includes situations in which domain name registrants have registered domain names that incorporate trademarks with the intent to benefit (e.g., by way of advertisements) from inadvertent traffic at the registrant's website.

In order to address cybersquatting, Congress enacted the Anticybersquatting Consumer Protection Act (ACPA). The ACPA identifies cybersquatting as:

(A) A person shall be liable in a civil action by the owner of a mark, including a personal name which is protected as a mark under this section, if, without regard to the goods or services of the parties, that person–

  (i) has a bad faith intent to profit from that mark, including a personal name which is protected as a mark under this section and
  (ii) registers, traffics in, or uses a domain name that–

   (I) in the case of a mark that is distinctive at the time of registration of the domain name, is identical or confusingly similar to that mark;
   (II) in the case of a famous mark that is famous at the time of registration of the domain name, is identical or confusingly similar to or dilutive of that mark; or
   (III) is a trademark, word, or name protected by reason of section 706 of title 18, United States Code, or section 220506 of title 36, United States Code.

15 USC § 1125(d)(1)(a).

Thus, in order for a trademark owner to bring a claim under the ACPA, the owner must establish: (1) that the mark is distinctive or famous; (2) the domain name registrant acted in bad faith by use of the mark; and (3) the domain name and the trademark are either identical or confusingly similar.

Around the same time that the ACPA was enacted, ICANN developed the Uniform Domain Name Dispute Resolution Policy (UDRP) to similarly address domain name abuse that impacts trademarks owners. The goal of the UDRP is to create a lower cost administrative process for the resolution of domain name disputes. In a UDRP proceeding, the trademark owner must prove the following three elements:

(1) that the domain name at issue is identical or confusingly similar to a trademark in which the complainant has rights;
(2) that the domain name registrant has no rights to or legitimate interest with respect to the domain name; and
(3) that the domain name has been registered and is being used in bad faith.

While a UDRP proceeding is beneficial in that it can result in a speedy and lower cost disposition of a domain dispute, the only available remedies in a

UDRP proceeding are the cancellation or transfer of the disputed domain name. As discussed above, whenever a registrants signs an agreement for registration of a generic or global TLD (gLTD) (e.g., .com, .net, .org, etc.), the registrant must agree to resolve any disputes with third parties regarding the domain name under the UDRP process.

Chapter 6
Intellectual Property Issues in Labeling and Marketing

Apart from whether a mark can be protected under the trademark law, there may be government regulations that can restrict or prohibit use of the mark in advertising or labeling any product(s) with the mark. Anyone using or selecting a trademark in the food industry should be aware of the nature and kind of regulations that may be applicable.

6.1 Governmental Controls Over Advertisements and Labeling

The production, marketing, distribution, and sale of food and beverage products in the United States are subject to a wide array government regulations on the federal and state levels.

The Federal Trade Commission ("FTC"), Food and Drug Administration ("FDA"), United States Department of Agriculture ("USDA"), and Alcohol and Tobacco Tax and Trade Bureau (TTB) share jurisdiction over claims made by manufacturers of food products. Since 1954, the FTC and the FDA have operated under a Memorandum of Understanding, under which the FTC has assumed primary responsibility for regulating food advertising, while FDA has taken primary responsibility for regulating food labeling.

6.1.1 Federal Trade Commission ("FTC")

The FTC is responsible for maintaining a competitive marketplace for both consumers and business and preventing unfair or deceptive trade practices. As such, it administers laws and regulations ranging from the content of clothing labels to laws requiring truth in advertising and prohibiting price fixing.

R.W. O'Donnell et al., *Intellectual Property in the Food Technology Industry*, 65
DOI: 10.1007/978-0-387-77389-6_6, © Springer Science+Business Media, LLC 2008

The FTC regulates food advertising under its statutory authority to prohibit deceptive acts or practices. The FTC Commission will find an advertisement deceptive (1) if it contains a representation or omission of fact; (2) that is likely to mislead consumers acting reasonably under the circumstances; and (3) that representation or omission is material.

The first step in the analysis is to identify representations made by an advertisement. A representation may be made expressly or implicitly. An express claim directly makes a representation of fact. An implied claim is not so straightforward and requires an examination of both the representation in order to determine the overall meaning or commercial impression of an advertisement. False claims can also stem from an omission of information that make an affirmative representation misleading. In other words, it can be deceptive for a seller to simply remain silent if such silence constitutes an implied, but false, representation.

The second step in identifying deception in an advertisement requires the Commission to consider the representation from the perspective of a consumer acting reasonably under the circumstances.

Finally, a representation must be material, i.e., likely to affect a consumer's choice or use of a product or service. Express claims involving health, safety, price, or efficacy are presumed material.

Any claim by an advertiser must have a reasonable basis. Where nutrient content or health claims are made, those claims should normally be substantiated by competent and reliable scientific evidence such as tests, analyses, research, studies, or other evidence conducted and evaluated in an objective manner by persons qualified to do so, using generally accepted scientific methods.

Accordingly, it would be deceptive for a food advertiser to make an express or implied nutrition or health benefit claim for a food unless, at the time the claim is made, the advertiser has a reasonable and substantiated basis for the claim.

The FDA has regulations that define certain absolute and comparative terms that can be used to characterize the level of a nutrient in a food. For example, "absolute" terms (e.g., "low", "high", "lean") describe the amount of nutrient in one serving of a food. "Relative" or comparative terms (e.g., "less", "reduced", "more") compare the amount of a nutrient in one food with the amount of the same nutrient in another food. The FTC has a policy of trying to harmonize its policing of deceptive advertising with FDA food labeling requirements and will normally defer to FDA standards. Therefore, use by an advertiser of FDA-defined terms in a manner inconsistent with FDA's definitions would likely be considered a deceptive act. By the same token, advertisers who comply with FDA nutrient and health regulations will not likely face action from the FTC in that regard.

6.1.2 Food and Drug Administration ("FDA")

The FDA is the federal agency broadly responsible for protecting the public health by assuring the safety and efficacy of human and veterinary drugs, biological products, medical devices, food, and cosmetics. In the area of foods, the FDA is responsible for ensuring that foods are safe and sanitary and that such products are truthfully and accurately presented to the public.[1] It accomplishes this by regulating food labeling, the safety of food products (except meat and poultry), and bottled water.

The Nutrition Labeling and Education Act (NLEA) directed the FDA to standardize and limit the terms permitted on labels, and allows only FDA-approved nutrient content claims and health claims to appear on food labels. The NLEA is designed in part to prevent deceptive and misleading claims on labels. At the same time, it is also intended that the nutrient content and health claims educate consumers in order to assist them in maintaining healthy dietary practices. The NLEA regulations apply only to domestic food shipped in interstate commerce and to food products offered for import into the United States.[2]

All product labels must have the following information (21 CFR §§ 101 and 105):

1. a statement of identity (common or usual name of the product);
2. a declaration of net quantity/weight of contents;
3. the name and place of business of the manufacturer, packer, or distributor;
4. list of ingredients, if fabricated from two or more ingredients. Each ingredient must be listed in descending order of predominance in the product by its common or usual name (21 CFR §§ 101.4 and 101.6).
5. Nutrition labeling unless exempted. See 21 CFR § 101.9.

The above requirements identify information that must appear on every label. The FDA has also detailed regulations that govern every aspect of labeling including the placement and typeface size of specific text. In addition, the name and form of the food may be subject to additional regulations as exemplified below.

---

[1] "Food" is defined in the Federal Food, Drug, and Cosmetic Act ("FDCA") to include "articles used for food or drink for man or other animals, chewing gum, and articles used for components of any such article."

[2] The labeling of food products exported to a foreign country must comply with the requirements of that country.

- The form of the food must be described in the label if the food is sold in different optional forms such as "sliced" and "unsliced", "whole" or "halves", etc.
- New foods that resemble traditional food and is a substitute for the traditional food must be labeled as an "imitation" if the new food contains less protein or a lesser amount of any essential vitamin or mineral.
- Beverages that purport to contain juice (fruit or vegetable juice), by way of label statements, pictures of fruits or vegetables on the label, or taste and appearance causing the consumer to expect juice in the beverage, must declare the percentage of juice.
- Beverages that are 100% juice may be called "juice". However, beverages that are diluted to less than 100% juice must have the word "juice" qualified with a term such as "beverage", "drink", or "cocktail". Alternatively, the product may be labeled with a name using the form "diluted ___ juice", (e.g., "diluted apple juice").
- Juices made from concentrate must be labeled with terms such as "from concentrate", or "reconstituted" as part of the name wherever it appears on the label.

Claims that characterize the level of a nutrient in the food (e.g., "low fat" or "high in oat bran") or make statements regarding the health benefits of food are subject to regulation. Terms such as "high", "lean", "extra lean", "high potency", "high", "rich in" "excellent source of", "good source of", "contains", "provides", "more", "added", "extra", or "plus" have specific requirements for use and those requirements may vary depending on the nutrient(s) in questions.

Only the terms that are defined in the FDA's regulations may be used if the food actually meets the FDA requirements for making the claim. For example, a claim that a food is "high" or a "good source" may only be made for a nutrient that has an FDA established daily value.

Claims relating to the health efficacy of a particular substance on a disease or related condition are also regulated. A manufacturer may only place health claims permitted by the FDA on labeling. Some of these are specifically authorized by regulation and include the relationships between calcium and osteoporosis, sodium and hypertension, etc. To the extent a health claim is not recognized by regulation, it may only be used with prior FDA approval.

The label of a food product may include the Universal Product Code (UPC) as well as a number of symbols which signify that: (1) a trademark is registered with the US Patent Office; (2) the literary and artistic content of the label is protected against infringement under the copyright laws of the United States; and (3) the food has been prepared and/or complies with dietary laws of certain religious groups. None of these symbols are required to be on the labels and are not under the authority of the FDA.

The FDA also regulates dietary supplements under a different set of regulations than those covering "conventional" foods. Under the Dietary Supplement Health and Education Act of 1994 (DSHEA), the dietary supplement manufacturer is responsible for ensuring that a dietary supplement is safe before it is marketed. The FDA is responsible for taking action against any unsafe dietary supplement product after it reaches the market. Generally, manufacturers of dietary supplements do not need to register their products with the FDA and are not required to obtain FDA approval before producing or selling dietary supplements.

The FDA's post-marketing responsibilities include monitoring safety (e.g., voluntary dietary supplement adverse event reporting) and product information, such as labeling, claims, package inserts, and accompanying literature. In terms of labeling, manufacturers must make sure that product label information is truthful and not misleading. Although dietary supplements require most of the same basic information statements as required with food, the FDA has special labeling requirements for dietary supplements that must be followed.

6.1.3 US Department of Agriculture ("USDA")

The USDA's Food and Safety Inspection Service (FSIS) is primarily responsible for ensuring that the nation's commercial supply of meat, poultry, and egg products is safe, wholesome, and correctly labeled and packaged. Although not identical, the FSIS labeling requirements are for the most part largely compatible with the FDA requirements.

A major difference with the FDA, however, is that all labels must be approved by the FSIS. However, the regulations permit generic approval of a final label that can be used without further authorization from FSIS. If the label is for a single ingredient amenable product that bears no special claims, guarantees, foreign language, or nutrition facts, it is a generic approval. If the label is for an amenable multi-ingredient standardized product and bears no special claims, guarantees, foreign language, or nutrition facts, the label can either be a generic approval or submitted to the Labeling Compliance Team (LCT) attached to a label application form. The label may have to receive sketch approval if it bears special claims (quality, nutrient content, health, negative, geographical origin, animal production, etc.), guarantees, foreign language, or nutrition facts.

6.1.4 Alcohol and Tobacco Tax and Trade Bureau (TTB)

The TTB, formerly known as the Bureau of Alcohol Tobacco and Firearms, is part of the US Department of Treasury and regulates the production, labeling, and advertising of distilled sprits, wine and malt beverages, including beer.

As with the FTC and FDA, the TTB has developed regulations governing labeling and advertising. Although there are specific labeling and advertising regulations for distilled spirits, wine and malt beverages, they share a common regulatory scheme in that they have common mandatory statements required on all labels and restrict claims regarding health benefits (e.g., "no hangover") and statements that mislead the public. These common requirements are as follows:

- The TTB must approve all labels for alcoholic beverages. Labels must include certain information, including: brand name, class and type, alcoholic content, net contents, name and address of bottler, country of origin in case of importation, etc.
- Label specifics with respect to size, placement, and legibility of required information as well as identifying information that must appear in English.
- The use of geographically significant terms is prohibited unless the product is produced in the place named or unless the term has lost its geographic significance through common use and knowledge.
- Positive health-related statements are prohibited unless pre-approved by the TTB on a case-by-case basis.
- The use of the following on labels or other matter accompanying the product is prohibited:

    o false or misleading statements;
    o statements disparaging competitor's product; and
    o obscene or indecent statements.

6.1.5 State Regulation

Individual states within the United States also regulate food labeling and advertisements. Many activities, but by no means all, relating to labeling are preempted by federal laws and regulations promulgated by the FTC, FDA, and other agencies. The strength of state power lies in each state's consumer protection laws which seek to prevent deceptive trade practices. Since the penalties for violating these laws can be severe in some cases, individual states can have a great influence on regulating advertising of food products.

6.1.6 Governmental Controls Outside of the United States

Regardless of one's knowledge of US labeling laws and regulations, these laws and regulations are only valid within the jurisdiction of the United

States. If a food or beverage is intended to be exported, the labeling requirements of the import country must be identified and followed.

For example, The European Union has general rules on the labeling, presentation, and advertising of foodstuffs, and the provision of the following particulars is compulsory on the labeling of foodstuffs:

- the name under which the foodstuff is sold;
- the list of ingredients, in descending order of weight;
- the quantity of certain ingredients or categories of ingredients;
- the net quantity of prepackaged foodstuffs expressed in metric units (liter, centiliter, milliliter, kilogram, or gram);
- the date of minimum durability in a specific format or the "use by" date for highly perishable foodstuffs;
- any special storage conditions or conditions of use;
- the name or business name and address of the manufacturer, packager, or vendor established within the European Community;
- the particulars of the place of origin which in case of the absence of such information might mislead the consumer;
- instructions for use;
- the actual alcoholic strength for beverages containing more than 1.2% alcohol by volume;
- a mark to identify the lot to which a foodstuff belongs; and
- treatments undergone, with specific indications for irradiated foods or deep-frozen foods.

6.2 Non-Governmental Controls

Many industries have trade associations that may have a code of conduct governing advertisements of their products. However, any such code of conduct is effectively non-binding on members. Moreover, if the code of conduct limits information provided to consumers (e.g., in the event of truthful competitive advertising) it could be considered anti-competitive and subject to challenge by the FTC. Examples of this would be an industry code of conduct that imposes a higher standard of substantiation for comparative claims than for unilateral claims.

Alternatively, false or misleading statements can be brought before the National Advertising Division ("NAD") of the Better Business Bureau (www.nadreview.org). In an action before the NAD, a competitor can file a complaint to take action against false or misleading statements, while avoiding the distractions and expense of full blown inter-party litigation or involvement with a government agency. After a complaint is initiated and

accepted, the NAD will investigate the matter, and if a deceptive or misleading practice is found, it will take action which can range from requesting
corrective advertising from the publisher, publication of the NADs findings,
or referral to a regulatory authority such as the FTC for further penalties.

6.3 Comparative Advertising

Comparative advertising refers to identifying a competitor's product by its
trademark and comparing it to the advertised product. Comparative advertising is widely used in the food industry to convey valuable information to
consumers. However, using such comparative advertising runs the risk that
a competitor could take action that the claims are either false or that the
advertisement infringes the competitor's trademark rights.

The FTC and courts have approved the use of brand comparisons where
the bases of the comparison are truthful, objective, and clearly identified. Advertisements containing truthful and non-deceptive statements that a product has certain desirable properties or qualities which a competing product
does not possess are permitted. However, false or misleading comparative
advertising can trigger liability for false advertising or trademark dilution.
Therefore, when a competitor's trademark is used in comparative advertising, the statements used must not have a tendency or capacity to confuse or
be considered to be false or deceptive.

In order to lessen the risk of litigation arising from comparative advertisement, the steps outlined below should be considered before making any
claims regarding a competitor's product.

- Only use the competitor's trademark to the extent necessary for comparison purposes.
- Use the competitor's trademark in the same way that the text is used
  throughout the advertisement. In other words, do not place emphasis on
  the competitor's trademark that would lead a consumer to believe that the
  trademark is associated with the advertiser's company.
- Do not disparage or mock the competitor or the competitor's trademark in
  the advertisement. Make sure any photographs of a competitor's product
  are accurate and do not place product in unfavorable light. In addition,
  do not use any photographs of a competitor's product that may be copyrighted by a competitor or a third party. Instead, use original photographs.
- Try to make sure your product's advertisements, packaging, and trade
  dress create a separate commercial impression from that of your competitor. In other words, a consumer looking at the two products or advertisements should not believe that the products are somehow related because
  of similarities in colors, fonts, stylization, etc.

- Include an easily visible disclaimer in any advertisement that uses the competitor's trademark. An example of such disclaimer is, "COMPETITOR'S TRADEMARK is a trademark of COMPETITOR. ADVERTISER does not make or license COMPETITOR'S TRADEMARK and is not affiliated or associated with Competitor."
- Verify that any statements regarding competitor or advertiser's product are true both explicitly and implicitly. The following steps should be taken:

  - Retain a certified, independent, and well-respected laboratory in the industry to perform the testing. Verify that the testing to be performed is the test generally accepted by technical or scientific community.
  - Make sure testing is fair and impartial. For example, ensure that the appropriate comparative products from the competitor's product line are compared.
  - All statements regarding a competitor's product should be supported by testing.
  - Make sure copies of the test methodology, test results, and samples tested are preserved in a safe place in case the tests have to be repeated at a later date.

Even if all the above suggestions are followed, there is nothing that can be done to prevent a party which feels it has been harmed from deciding to bring a court action to protect its rights. It is recommended to consult legal counsel regarding individual advertising claims before a comparative advertising campaign is launched.

Part II
Implementing IP Practices and Procedures

Chapter 7
Seven Basic Steps to Getting Started

This chapter will review the actions that a company should consider in order to protect its intellectual property assets. The following example will be used throughout this chapter to discuss the implementation of basic intellectual property practices.

> *Example*: Tastewell Industries is developing a new cheese product. The product is a mix of certain processed cheeses and various fruits. Tastewell wants to develop the product as well as a marketing plan. It has assigned two of its senior food scientists the task of developing and testing sample products that could meet this new market niche. Several younger scientists and analysts in the R&D group will also be involved in the analysis and testing. It is also necessary to involve a long standing Tastewell vendor for the cheese flavor, as well as a new vendor for supplying the fruit ingredient. The marketing team will include several Tastewell employees as well as an outside marketing firm that Tastewell has retained to assist in developing the marketing package. Finally, Tastewell has decided that it will approach BigFoods Distribution for a nationwide launch of its products.
>
> Tastewell's arch-rival, Bland Foods, Inc. has a history of copying Tastewell's products, and on occasion has attempted to solicit Tastewell employees to change positions so that Bland Foods can get the inside track on the latest Tastewell developments.

7.1 Confidential Disclosure or Non-Disclosure Agreements

Confidential Disclosure Agreements, which are also referred to as Non-Disclosure Agreements or NDAs, refer to a contract that protects confidential or trade secret information ("Confidential Information") from disclosure to

R.W. O'Donnell et al., *Intellectual Property in the Food Technology Industry*,
DOI: 10.1007/978-0-387-77389-6_7, © Springer Science+Business Media, LLC 2008

third parties. NDAs are commonly included as a part of an employment contract, and are also included in, or form, a separate agreement with vendors, contractors, and sometimes customers.

An NDA should include the following provisions:

(a) A clear definition of the information that is to be held in confidence, as well as a provision defining how the information should be marked or identified, for example with a "Confidential" stamp or label. Employment agreements typically adopt broader definitions requiring employees to maintain any work-related information that the employee develops or has access to as confidential. Supplemental agreements for specific employee development projects may also be used to more clearly identify the information that is subject to the agreement.

(b) A specific recitation of the limited purposes for which the Confidential Information can be used. For example, in an NDA with an outside vendor, the Confidential Information may be provided for specific testing and evaluation of a product, such as a flavor preference study. For an advertising or marketing agency, the Confidential Information could be limited to the specific purpose of package development, branding, or other specific marketing tasks for the benefit of the disclosing party.

(c) A recitation that the receiving party cannot breach the confidential relationship, induce others to breach it, or induce others to acquire the Confidential Information by improper means.

(d) A recitation of exceptions to the confidentiality requirements. This generally will exclude any information that: (1) the receiving party can show they were already in possession of at the time of the disclosure; (2) information that is in the public domain through no fault of the receiving party; and (3) information that the receiving party receives without restriction from a third party.

(e) The time period that the information must be held in confidence. This can be any reasonable term agreed upon by the parties, and often falls in the range of 2–5 years, depending on the technology in question and the disclosure's purpose. For example, an advertising agency that is working on an advertising campaign that is going to be released within a year would not need an NDA term that extends beyond the advertising campaign release. However, for an outside consultant that does taste testing of beta products to determine a preferred version of a product for commercialization, the NDA could justifiably require that the information be held in secrecy for 5 years or more.

(f) A provision defining the remedy available to the non-breaching party in the event of a breach or impending breach. This should recite that the disclosing party is entitled to injunctive relief for breach of contract by the receiving party to prevent the release of the Confidential Information. It is also possible to include a liquidated damages provision as an incentive for the receiving party not to disclose the Confidential Information, although this is less typical given the circumstances surrounding most NDAs where the disclosing party is attempting to obtain information or services from the receiving party.

(g) Other miscellaneous provisions can include: (1) a provision for return or destruction of the Confidential Information when the task or review by the receiving party is completed; (2) a transfer of ownership of any additional intellectual property that results directly from the Confidential Information and/or in connection with the work being done by the receiving party; and (3) a recitation of the courts or jurisdictions where any potential dispute will be resolved.

While NDAs have several advantages, there are a few drawbacks that they cannot address. Many large companies will not sign NDAs for any outside submissions, regardless of purpose. In fact, some large corporations require the opposite: a signed statement saying that no information will be held in confidence and that the party submitting the information will rely exclusively on any intellectual property rights (such as patent or trademark rights) that they applied for or may obtain as the sole recourse against the receiving party in the event that their information is used. This typically occurs as a result of a competitor's parallel development efforts. A competitor may accuse a company receiving its confidential information of theft or misappropriation even though the receiving company was already working on a similar development. Rather than face potential law suits or bad publicity, the policy of such express waivers of confidentiality shields the receiving company.

Even if an NDA is signed, if the Confidential Information is purposefully or even inadvertently disclosed, it may not be possible to "put the genie back in the bottle." Although damages may be available against the discloser, an actual public disclosure cannot be undone. This could result in inadvertent loss of the ability to seek patent protection in many countries that have absolute novelty requirements. While the United States allows a 1 year grace period from the date of first public use or disclosure of an invention, if the owner of the Confidential Information is unaware of the disclosure, US Patent rights could also be lost. Furthermore, any trade secret information that is publicly disclosed ceases to be a trade secret.

Example: In the scenario presented, Tastewell should execute employment agreements with all employees who will have access to the project, including its senior food scientists as well as anyone that will be working on the new product, its marketing employees, and any others who will have access to the information. If confidentiality terms were not included in the employment agreements, it can enter separate agreements, preferably at the same time as an annual review and raise so that there is no question regarding a potential lack of consideration for the new NDA provisions. The confidentiality term should extend for a time period beyond the end of employment, especially if a rival such as Bland Foods is known to poach employees in order to copy products or product concepts. In the event that an employee with knowledge and/or Confidential Information related to the new cheese product leaves Tastewell, the NDA should provide for injunctive relief to prevent the Confidential Information from being improperly disclosed.

In addition, any outside vendors, whether for the cheese flavoring or for marketing, should sign a project specific NDA with Tastewell. The vendor agreements should also have specific terms spelling out ownership of any new developments made using the Confidential Information while carrying out the work for Tastewell.[1]

BigFoods Distribution will likely require advance notice of the product. If Tastewell learns that BigFoods will not sign an NDA, and to the contrary requires a specific waiver of confidentiality to accompany any offer or disclosure, Tastewell should have any patent applications for its new product filed with the US Patent and Trademark Office before any disclosure. This will protect any potential US or foreign patent rights that Tastewell may choose to pursue.

7.2 Assignment of Rights

An Assignment is a contract between two parties in which the rights owned by one party are assigned to the other party. In the United States, all rights to an invention are initially vested in the inventor(s). Accordingly, a formal Assignment is important to transfer rights from an inventor or inventors to the company. For employees, the obligation to assign inventions made in the course of an employee's regular job duties can also be included in the employment contract. If the obligation to assign inventions is not included in

---

[1] A sample NDA is set forth in Appendix A.

an employee's employment agreement, a separate agreement can be entered including the obligation to assign inventions to the company. This separate agreement should preferably be made and signed with the employee in connection with an annual review and raise so that there is no issue with respect to consideration for the agreement.[2]

Many companies offer employee incentive programs that pay bonuses for making inventions that help the company. This typically takes the form of a lump sum payment at the time a patent application is filed or a patent is granted. Some countries, such as Germany, have specific statutory requirements that define the amount that an employee must be compensated for any invention that is used by the company.

When working with third parties, it is common to include assignment terms in the vendor contract for inventions related to the specific work. For patentable inventions, this can be critical if a breakthrough is made by the vendor in connection with the development of a company's product. If no agreement is reached before the contract work is undertaken, rights would belong to the vendor or vendor's employee, creating the potential for being forced into a sole source of supply, or worse, having the product or flavor which the company paid the vendor to develop offered to third parties without any ability for the company to control it.

> *Example*: Assignments are critical for any copyrightable subject matter, for example, as might be created by Tastewell's outside marketing firm in connection with marketing the new cheese product. In the absence of a written assignment for works made under contract for Tastewell, the copyrights would be owned by Tastewell's outside marketing firm. For assignment of copyrights, the assignment should clearly recite that any work done by the outside marketing firm was done as a "work for hire" and that all copyrights are owned by Tastewell.

7.3 Employee Education

Employee education is one of the cornerstones of a successful intellectual property program. Inadvertent disclosure of confidential information or new inventions that are being developed can easily result in any potential

---

[2]A sample Assignment contract from an individual inventor to a company is set forth in Appendix B.

intellectual property rights being lost, or worse, landing directly in a competitor's hands. The only way to effectively address this is to train employees such that they understand the ramifications of their actions and the potential cost to the company in terms of lost intellectual property rights and lost profits on new products or developments that cannot be protected.

Many companies hold intellectual property seminars that are presented by in-house or outside intellectual property counsel. These programs educate employees on the basic tenants of intellectual property law in a practical setting, and how they can handle particular situations. A typical program would include a review of what may constitute patentable subject matter given the company's technology field, the potential bars to patentability that an employee should be aware of, and a review of the company's system for documenting inventions and subsequently handling those documents. Employees that regularly deal with third party vendors should also receive special training related to risks associated with dealing with vendors.

7.4 Accurate Record Keeping

7.4.1 Patents

While the United States is currently a "first to invent" country, meaning that the first true inventor is entitled to any patent rights if more than one inventor files a patent application for the same invention, legislation is currently being considered to change the US Patent system to a "first inventor to file" system.

Under the current "first to invent" system, it is important for inventors to document when they first conceived of the invention, as well as all of the activities that were carried out in actually "reducing the invention to practice," either by making or carrying out the invention, or by filing a patent application for the invention, referred to as a "constructive" reduction to practice.[3] This information should be documented in written form, and should be dated and witnessed by another person. A preferred form for documenting this type of information is an inventor's notebook, which is a bound book with numbered pages where information can be recorded. Each page includes a space where it can be dated and witnessed once the information has been entered. Completed notebooks should be kept by the company for reference, if needed, since it may be years before the information is actually needed.[4]

---

[3]"Reduction to practice" is discussed in Chapter 1.

[4]A sample inventor's notebook page is set forth in Appendix C.

Another document that the company should have is an employee invention submission form. This should include sections for a complete description of the invention, as well as any potentially critical events that occur, such as a disclosure to others. This form serves two purposes: (1) it can serve as the vehicle for in-house review and a determination of whether patent protection is going to be pursued; and (2) if the company proceeds with a patent application, it can act as the vehicle for transmitting information on the invention to patent counsel for searching and/or the preparation of a patent application.[5]

For each invention disclosure that is ultimately pursued as a patent application, a company should open a separate file as a place to store all information related to the invention, including copies of the relevant completed pages of the inventor's notebook, the invention disclosure form, any known prior art that might relate to the invention, as well as all correspondence and documents related to the preparation, filing, and prosecution of a patent application before the USPTO.

7.4.2 Trade Secrets

Trade secret protection depends on defining and following a strict set of policies for handling the information that is being protected as a trade secret. Trade secret policies should be documented in a policy manual, and all employees that have access to the trade secret information should receive training on handling trade secret information and should be required to periodically review the company's policy regarding treatment of such information. Policies should include how to identify and mark trade secret information, including information or technology under development. Identification can be as broad as "all information related to project X" or down to a specific formula for a product, such as a flavor ingredient or mix, and can be set by management or counsel. Marking should be on both physical and electronic documents, using labels such as "Confidential," "Secret," "Trade Secret," or "Proprietary Information" of the company.

Access to the trade secret information should be on a need-to-know basis inside the company. Access to the information should be on a "log in–log out" basis, whether on paper or electronically. Any electronically stored or transmitted information should be encrypted based on a defined procedure. A policy should also define storage of the information when an employee is away from his work area, and may include a locked central or private storage

---

[5] A sample invention disclosure form is set forth Appendix D.

area. The reason for this high level of security is that in a misappropriation lawsuit, a court's inquiry will not only focus on the bad acts of the accused party, but will also examine and consider whether the company asserting its trade secret rights adequately maintained and safeguarded its trade secrets.

Example: In the present case, Tastewell may consider trade secret protection for its cheese flavor and its product formula. Tastewell not only needs to establish and document its trade secret policies, but it must also inform its need to employees of these policies and implement a plan for carrying them out. Any Tastewell vendor that will share Tastewell information should also be required to sign an NDA with an extended or unlimited term.

7.5 Patent and Trademark Searches

7.5.1 Patents

There are a number of different patent searches that can be useful for a number of different purposes, including patentability, infringement clearance, validity, and state-of-the-art searches.

A patentability search is the most basic search, and involves searching US patents and patent publications as well as potentially other patent and non-patent literature to determine whether an "invention" meets the USPTO requirements for patentability, based on the documents identified by the search. This is a useful tool to gauge the potential for patentability; however, it is not a guarantee that a patent would ultimately be granted. The limitations on patentability searches are that they are generally not exhaustive, and other more pertinent references may ultimately be identified from areas not searched. Accordingly, while negative results can be relied upon, a positive search report merely leaves the possibility of patent protection open.

An infringement clearance search is a search of US patents that remain in force for potential infringement by a new product that is being developed. Infringement clearance searches should be performed prior to the product's launch. This type of search involves a review of the patent claims that remain in force in the relevant classifications for the product being developed. As the search requires a specific review of each independent claim of the relevant patents, it is more complete. If done early enough in the design or development process, a clearance search allows the new product to be modified before potential infringement.

A validity search is a prior art search of US and foreign patent and non-patent documents that is directed against the claims of a specific patent. This can be used to determine whether the claims of a known or asserted competitor patent are valid. The results of the search can be used for negotiations, or can be used to invalidate the patent in a court or USPTO proceeding.

A state-of-the-art search looks at US and possible foreign patent document collections for representative technology and developments in a particular field. This is used as a research tool for resolving a particular problem or for examining the type of work competitors have done in a given field.

7.5.2 Trademarks

Trademark searches determine whether a company can adopt a trademark for its product or service. A trademark search can be done in the Federal Trademark database, or can be done in one or more comprehensive databases. Trademark searches should be carried out before adopting a trademark to determine whether any third party has used the trademark for the same or similar types of goods, and may therefore have superior rights in the trademark. Adopting a third party's mark, whether knowingly or unknowingly, may result in a lawsuit for trademark infringement. A trademark search can help avoid this expensive litigation.

7.6 Decide on the Type of Protection Early in the Inventive Process

The earlier a company decides what type of intellectual property to pursue, the lower the likelihood that rights will be inadvertently lost. For patents, it is important to observe specific timelines before the invention's first public disclosure, use, or offer for sale. These dates will also determine certain statutory bars to patentability in the United States. For trade secrets, the earlier that a decision is made to protect a new product or even its method of manufacture as a trade secret, the easier it will be to ensure that some information is not inadvertently released.

7.7 Speak to an Intellectual Property Attorney

As the facts and circumstances surrounding product development and branding vary widely, consulting with an intellectual property attorney is highly recommended.

Example: During development of the new product, one person on the marketing team leaves Tastewell for Bland Foods. While the former marketing employee does not know the specifics of the new Tastewell product, he has sufficient information to allow Bland Foods to get a head start on a competing product.

In this situation, if the marketing team employee had signed a project-specific NDA for the new Tastewell cheese product, Tastewell could enforce the NDA against the former employee and Bland Foods. Moreover, in such a situation, Tastewell's internal procedures for handling trade secret information could also have prevented access to sensitive information if it was handled properly from the beginning of the project. Finally, having a pending patent application for the product as well as any specifically developed processes required to make the product, and an intent-to-use trademark application for the new product name would provide a further line of defense against Bland Foods.

Chapter 8
Deciding Between Patent or Trade Secret Protection

Many companies use both patents and trade secrets to protect inventions and often face a choice between the two forms of protection. Each has advantages over the other and should be carefully considered when forming the appropriate intellectual property strategy.

A first consideration is the difference in subject matter between trade secrets and patents. Trade secret protection covers a wider range of possible innovations and inventions and it has a potentially unlimited duration. The Uniform Trade Secret Act ("UTSA") defines trade secrets as any information, including a formula, pattern, compilation, program, device, method, technique, or process that derives independent economic value from being "secret".[1] While concepts, databases, and compilations are generally not patentable, the UTSA expressly protects them if they are valuable to the business and the business takes steps to keep them secret. Further, other categories of information that can be subject to trade secret protection where patents would not normally apply include customer lists, product pricing, strategic planning, company policies, market analyses, etc.

A second consideration is the different types of litigation remedies. A US patent grants the owner the right to exclude others from making, using, or selling the invention throughout the United States. In return for this right, the patentee must disclose to the public how to make and use the invention. Thus, even a competitor who independently, and without any knowledge, develops or reverse engineers an invention that is covered by a patent cannot practice the invention without infringing the patent. In contrast, if a person that is privy to the trade secret unlawfully uses or discloses the trade secret, its owner can enforce the trade secret by filing a suit for misappropriation. Trade secrets are litigated less frequently than patents because the owner

---

[1]The UTSA is discussed above in Chapter 2.

R.W. O'Donnell et al., *Intellectual Property in the Food Technology Industry*,
DOI: 10.1007/978-0-387-77389-6_8, © Springer Science+Business Media, LLC 2008

may not want to disclose the trade secret as part of the litigation discovery process or in court proceedings.

A third consideration concerns the process for creating rights and the term of those rights. Unlike patents, there are no formal application or registration requirements for trade secrets. Any valuable and secret information used by a business is protected, as long as the business takes reasonable steps to keep it secret. This generally means the initial cost of trade secrets is lower. However, sometimes the cost of enforcing and updating procedures to keep information secret is more than the cost to secure a patent.

The following table summarizes these and other differences between patents and trade secrets.

	Patent	Trade secret
What is the protected subject matter?	Inventions (e.g., processes, machines, manufactures, compositions of matter, improvements of the foregoing, etc.).	Business information which gives the owner a competitive advantage over competitors and is maintained in secret (e.g., formulas, patterns, compilations, programs, devices, methods, techniques, processes, etc.)
What is the term?	20 years from patent application filing date.	Indefinite, as long as information is kept secret and used in the business
How is it acquired?	Filing a patent application with the USPTO.	Acquired upon creation. No formal application process
What are the requirements?	The invention must be patentable subject matter, useful, novel, and non-obvious	The information must give the owner a competitive advantage and must be maintained as a secret
Anticipated costs?	Patent application filing fee, patent issue fee, post-allowance maintenance fees, and attorney time required to prepare and prosecute a patent application	No specific costs. However, costs are typically incurred in trying to maintain the subject matter as a secret (i.e., confidentiality and non-disclosure agreements, implementing internal policies for treatment of confidential information, etc.)

	Patent	Trade secret
Can others use the invention/ information?	Others cannot practice the claimed invention without permission (i.e., license). However, others may design around the invention	Trade secrets can lawfully be reverse engineered by others. In addition, others may use the information pursuant to a non-disclosure or confidential agreement
Can the rights be lost?	Patent rights can be terminated if the validity of the patent is challenged. In addition, rights to a patent can expire if the invention was in public use or sold or offered for sale more than 1 year prior to filing a patent application (typically through the inventors own actions)	A trade secret is extinguished if it is disclosed by the owner or anyone. For example, if an owner of a trade secret files a patent application for the invention, the trade secret is lost upon publication of the patent application or issuance of the patent
How is it enforced?	Assert rights against others for patent infringement	Assert rights against others for misappropriation of trade secret

Patent protection is typically favored when:

- it is likely that a product can be reverse engineered;
- the innovation might be discovered by others simultaneously;
- the technology is difficult or expensive to be kept secret;
- the technology must be disclosed to be of use;
- the subject matter is patentable; and
- the commercial value of the innovation exceeds the registration and maintenance costs.

Trade secret protection is typically favored when:

- the subject matter is unpatentable;
- the subject matter is part of a relatively "crowded" art;
- the potential profits are low;
- keeping the innovation a secret is easy;
- the potential market is likely to last longer than 20 years;
- the technology is developing rapidly and the innovation is likely to be obsolete in a few years; and
- an invention is no longer patentable.

Example: Tastewell's new fruit and cheese product qualifies for both patent and trade secret protection. Although the product is completely novel, once it is sold or in use, it can easily be analyzed and duplicated. Tastewell inquires as towhich form of protection is best suited for the invention.

In this scenario, trade secret protection is available only during the research and development phase of the product. Once the product is complete, patent protection is preferred as it will give Tastewell the right to prevent others from duplicating the commercially available product. It is noted that Tastewell should file its patent application as early as possible after completion of the invention, and must file the application within 1 year after any sale or public disclosure.

On the other hand, suppose the process for making Tastewell's cheese and fruit product is novel and qualifies for both patent and trade secret protection. Examining the finished product alone would make it difficult, if not impossible, to duplicate the process. In this scenario, a trade secret may provide adequate protection since the trade secret is not revealed in the commercial product. Furthermore, the trade secret has an indefinite lifetime as long as it remains secret.

Chapter 9
Developing and Managing an Intellectual Property Portfolio

It is important to have a well-organized and focused intellectual property management program in place in order properly develop and enforce a company's intellectual property rights. Factors that should be considered in developing an intellectual property portfolio management programs include: (1) strategic considerations in developing an IP portfolio; (2) administrative issues associated with managing the IP portfolio; and (3) ongoing IP diligence protecting rights and pursuing others.

9.1 Developing an Intellectual Property Portfolio Strategy

Developing a company's IP portfolio strategy requires four steps: (1) identification of existing IP assets (an IP audit); (2) determining which of those assets are "core" assets (i.e., those assets that have a strategic importance to the company); (3) allocating resources to core and non-core assets as appropriate; and (4) setting up a program to ensure that those assets are periodically reviewed and maintained to ensure that the IP portfolio is developed consistently with the corporate business plan.

9.1.1 Identification of IP Assets

In order to properly identify IP assets, it is critical that a thorough IP audit be performed by the inventors in coordination with an experienced IP attorney. The IP audit should identify copyrights, trademarks, trade dress, trade secrets, confidential information, copyrights, mask works, domain names, industrial designs, patents, patentable inventions, and other IP assets owned by the company. Some registered assets (e.g., patents, trademarks, and copyrights), require periodic maintenance in the form of payment of official fees and filing of official documents. If these fees are not paid or the official documents are not timely filed, these rights can expire. Consolidating all of this information in a single location can help ensure that all of a company's intellectual property can be easily identified, tracked, and maintained.

Along with intangible assets, an intellectual property audit should identify any encumbrances on the company's IP assets. These encumbrances include agreements (such as distribution agreements, licenses, software licenses, franchise agreements, assignments, covenants not to compete, employment agreements, third party development agreements, and liens against the intangible assets) and any other liabilities that a company may have with third parties. The audit should identify whether or not these agreements affect the company's IP or other intangible assets.

9.1.2 Determining Whether the Identified IP Assets Are Core Assets

Following IP asset identification, the assets must be evaluated against the company's business plan and current financial condition to determine whether the assets are core assets. A core asset is one that is critical to the success of the company. This critical evaluation is important because IP assets can be expensive to secure and maintain. Thus, IP strategy is often a balance between long-term strategies and a company's economic resources.

The following factors help to determine whether IP assets are core assets and important to the success of a company:

- Are the IP assets currently used and will they be used in the future, or are they related to a product or a part of the business that is obsolete?
- Do the IP assets support a program (either marketing, legal, or research and development) that provides the corporation with a competitive advantage?
- if the IP assets were lost, would the company be negatively impacted?
- Do competitors have IP assets that will negatively impact the company in the marketplace?
- Does the company need additional IP assets in order to cover the important aspects of the company's business plan?

9.1.3 Properly Allocating Corporate Resources to Core and Non-Core IP Assets

Because there are many different types of IP assets, it is difficult to generalize the scope of protection for each type of IP. However, each type of IP typically has multiple levels of protection and these multiple levels of protection, in turn, have varying costs. Core assets should be given the broadest scope of protection since they are critically important to the corporation. Non-core assets may be given minimal protection; or even none at all. It is important to note that corporate resources include not only financial resources, but also human resources and executive attention.

For example, it may cost upward of $10,000 to secure a utility patent in the United States. In addition, if there is a potential to use or sell the invention on an international level, patent protection may be required in individual foreign countries. Patent protection in foreign jurisdictions can be more expensive than the United States. (e.g., filing a European patent application can be nearly twice as expensive as a corresponding US filing). One way to mitigate costs may be to file regular (non-provisional) utility patent applications for core IP assets and provisional patent applications for non-core IP assets. As described in the patent chapter, provisional patent protection provides a vehicle for a company to keep its options open for a 1 year period to decide whether to pursue a regular patent application. At the cost of approximately $2,000–3,000, this can be a minimal financial commitment for a company, although more costs will be incurred if the company decides to file a regular (non-provisional) patent application.

The degree to which corporate resources are devoted to certain IP assets depends largely upon the value of the IP to the company. If those assets provide minimal value, then a company may choose not to put a large amount of resources into defending the IP. An IP attorney can help correlate core IP assets with a company's goals in order to properly allocate the appropriate resources.

9.1.4 Setting up a Program for Periodic Review of IP Assets

Following the initial IP asset review, it is important to ensure that those assets are periodically reviewed. These successive reviews ensure that existing IP assets are properly maintained and an informed decision is made to retain, sell, or dispose of non-used assets.

The frequency with which such reviews are conducted will vary greatly depending on whether the company is a startup, an emerging entity, or a long established corporation. At the very least, a yearly audit should be conducted. Preferably, quarterly reviews of a current portfolio should be undertaken. The amount of resources to devote to the review will also depend upon the importance of the IP assets to the company. If the only core asset that has been identified is the corporate name and that has been protected with a trademark registration, in-depth IP audits on a yearly basis may not be necessary.

9.2 Administrative Issues for Long-Term IP Portfolio Management

After building an IP portfolio, administering the IP portfolio is a responsibility that requires integration of knowledge from many different parts of an organization. A well-executed IP portfolio management program must be closely tied to a company's strategic plan, marketing initiative, legal department, and corporate research and development initiatives. Although tying all these aspects together in a comprehensive IP portfolio management program may appear to be daunting, there are a number of steps that make the management of an IP portfolio a much easier endeavor, regardless of the size of the IP portfolio.

9.2.1 Docket Management

Obtaining and maintaining intellectual property rights is a deadline-driven process. If a company does not meet its IP-related deadlines, the consequences can range between payment of late fees or penalties to a complete loss of the IP rights. Accurately tracking US and international deadlines ensures that formal requirements are fulfilled. Such tracking is the primary responsibility of an IP portfolio docketing system maintained by the company or its IP counsel.

Docketing software is specifically tailored for IP portfolio management to intake relevant information needed to properly track IP deadlines. The software elicits required information and calculates the relevant deadlines for both US cases and all foreign counterparts. The software also provides reminders at certain intervals as the deadlines approach, and these reminders can be automatically e-mailed to the appropriate individuals. Such reminders can be tailored to provide not only information regarding the particular IP asset, but also information related to an organization's procedure.

For example, rather than taking the information generated by the docketing software, inserting this information into a memo and sending it to the person within the organization who will be responsible for making the decision regarding whether or not to proceed with complying with a deadline (which may include the payment of significant fees), automatic notices which include the IP asset at issue as well as the specific steps that are required to be taken within the organization can be automatically provided by IP docketing software. Thus, the software can be specifically tailored to the procedures of an organization in order to make management of the IP portfolio more efficient.

9.3 Ongoing IP Diligence: Protecting Rights and Pursuing Others

9.3.1 Defending Your IP

Aggressively pursuing competitors with litigation can be reckless if the company is vulnerable to an easy counterattack. Pursuing a solid defensive strategy means putting into place the minimum measures in order to protect core IP assets.

IP protection is, by its very nature, defensive. Patents prohibit others from using the products or processes generated from research and development programs. For a technology-centric company, this can represent the core of the business. Trademarks and service marks protect one of the most important aspects of a corporation's image; its name, logos, etc. Significant amounts of marketing and advertising budgets are typically allocated to using trademarks and service marks and, therefore, it is extremely important to protect these valuable assets. Copyrights protect a corporation's expression of ideas such as drawings, websites, marketing materials, documents, and software. Trade secrets, if handled properly, help protect customer lists, business plans, and strategic plans. Since large amounts of a company's resources are typically allocated to all of these endeavors, it is important to consider protection for all aspects of a company's IP portfolio.

Pursuing a vigorous defensive strategy has additional benefits. Such a strategy not only ensures that the IP rights are protected, but can also strengthen those IP rights. For example, with effective trademark protection, if a trademark holder does not provide notice to competitors or the public at large from using a trademark to refer to a particular product (for example, using the term Kleenex as opposed to tissue), the trademark could ultimately become "genericized" whereby anyone can use the mark. Thus, rights to the mark will be lost and it will cease to become an asset for the company. This happened with respect to the term "aspirin" for the Bayer Corporation.

A similar defensive strategy can be pursued using patent protection. If a corporation has a product that is protected by a patent, it can defend against another company trying to copy that product by asserting the patent against the infringer. If successfully implemented, such a strategy will protect the product and, as a result, the company's market share.

There are many different reasons why companies initiate intellectual property litigation. One of the primary defensive weapons that a company may have to counter an intellectual property dispute is its own patent portfolio. When a company that is aggressively pursuing patent infringement litigation realizes that the alleged infringer is not going to "cave in," the company may become more reasonable regarding its demands. Without a defensive patent portfolio as a counter-threat, a company can become vulnerable to repeat attacks from different competitors. Often, entire industries will be cross-licensed in such a fashion. However, such cross-licenses are only granted to those companies that have an IP portfolio to bring to the table.

There are several measures that should be considered when putting together a strong defense:

(1) Are all core assets are protected?

It is critically important to ensure that all IP assets that have been identified as core assets are properly protected. If they are not protected, a company may be vulnerable to a competitor entering the market with a similar competing product and quickly eroding the company's market share or operating margins.

Having IP assets on fundamental technology will help insulate a company from attacks by competitors having IP assets on related technology. Although such claims may be frivolous, defending against these attacks can drain smaller companies of desperately needed capital and distract management from its primary focus.

(2) Have clearance opinions been sought on all products to ensure that a company is not charged with infringement of competitors' IP portfolios; particularly patents or trademarks?

Just as critical as protecting a company's core assets, is ensuring that its products do not infringe competitor's IP rights. One way to determine this is to periodically review the industry for competitors' IP assets. If relevant IP assets are identified, it is recommended that a company seek a clearance opinion or a freedom-to-operate opinion from an experienced IP attorney. This can help ensure that the company's invention is free to be used or sold without charges of infringement by competitors.

(3) Have vulnerabilities or gaps in competitor's product lines or IP assets been identified?

Once a company assesses its own vulnerabilities, it is critical to identify those of its competitors. Although this does not require immediate or offensive action, such information will become invaluable should a competitor begin to make demands to the company.

(4) If you have determined not to pursue patent protection for an invention, have you considered a defensive publication in order to keep a competitor from gaining right to the idea?

As discussed above, a company will need to make strategic decisions about which IP rights to expend resources for protection. If a decision is made to forego protection for a particular asset, a company should consider whether the asset could provide competitors with an advantage if independently developed. In order to keep competitors from gaining such a strategic advantage, publication of an article can dedicate information about the invention to the public in general and mitigate the harm resulting from a competitor's use of the invention.

9.3.2 Leveraging IP Rights

An IP portfolio is a significant strategic asset that can be used to leverage favorable outcomes against competitors. A primary offensive weapon is the threat of an infringement suit. A company can also use licensing and the threat of (or actual) litigation in order to generate revenue from its IP assets.

For example, during research and development (R&D) a company may cover various aspects of a potential product. As the product comes to fruition, it is often apparent that alternatives that were pursued during R&D and protected with IP assets might be downgraded to non-core assets. In this manner, a company may wish to use the IP covering "core" assets to keep other competitors out of their marketplace. However, alternative designs or processes that did not end up in a product that is strategically important to a company, or IP assets that cover products that were ultimately not pursued due to a company's strategic plan, may be available for licensing to third parties. Therefore, a company may gain a revenue stream without hurting its core market.

When deciding to aggressively pursue an offensive IP strategy, corporate management must be willing to back up its any infringement allegations. There are several measures that should be considered when putting together a strong offense:

1. Are competitors litigation-averse (i.e., do they have a history of settling litigations quickly)?
2. Are your corporate management and board of directors willing to use the threat of litigation (or actual litigation) in order to enforce the IP rights or gain a strategic advantage over your competitors?
3. Is the company willing to monetize its unused IP assets through licensing, franchising, or sale?
4. Is the company willing to develop IP assets solely for the purpose of monetizing them?

5. Is the company willing to search for and purchase IP assets and other corporate assets from those that are available in the marketplace?

All these measures should be considered carefully. A company should carefully weigh the risks and benefits prior to embarking upon an IP strategy that incorporates an aggressive offense as a central part of its strategy. Such a strategy will only be successful if it has the full support of management and directors, as it will take up large portions of corporate resources.

Chapter 10
Enforcement and Infringement of Intellectual Property Rights

10.1 Policing Your IP Rights

After a company invests time and resources in developing and protecting its intellectual property rights, it is important to police the market to ensure that no competitor is improperly benefiting from such intellectual property investments. Policing of intellectual property rights can be broken down into a four-step process.

1. First, the rights must be identified by class: trademark, trade secret, copyright, or patent. It is also important to remember that any single right may be protected by more than one type of intellectual property protection. For example, a patentable or trade secret right may have an associated trademark, or there may be copyrightable materials that accompany distribution of the patented item. Copyrighted instructions may be an important element in the commercialization of a trade secret. Consider the sale of a trade secret formula where the temperature, rate of mixing, and amount of the intermediary is critical to its performance in the end product. While the owner of such information would not want the disclosure to reveal the trade secret, the copyrighted instructions could be an important sales tool because it may be the only way to make the formula commercially useful to your customer. This identification process should be conducted carefully to consider all possibilities for protection.

2. Second, the protection of intellectual property rights identified through prior analysis will require consideration of the competing marketplace of interest. If the identification step yields a product having commercial interest that is potentially patentable and marketing has identified or coined a trademark for the product, there are competing interests associated with the different forms of protection. On one hand, the patent interest requires diligence to be sure that the invention is not disclosed or sold more than 1 year prior to application of the patent. On the other hand, trademark rights

R.W. O'Donnell et al., *Intellectual Property in the Food Technology Industry*,
DOI: 10.1007/978-0-387-77389-6_10, © Springer Science+Business Media, LLC 2008

are based on use and there is a desire to initiate commercial use of the mark in association with a product as soon as possible. As noted in the trademark chapter of this book, it is possible to file an "Intent to Use" trademark ("ITU") application before there is any disclosure of the invention or use of the trademark. There is a similar procedure, known as a provisional patent application that can be filed to preserve a patent application filing date prior to any disclosure. Use of these two vehicles permits the preservation of patent and trademark rights while efforts are undertaken to gauge the market's commercial interest. This relatively simple solution to the potential problem is made possible through diligence in the identification of the concerned rights and remaining mindful of the critical first date of the underlying right.

3. Third, vigilance in the marketplace is a critical component of intellectual property management. Market vigilance is the key to gathering information about competitive practices and products. This information is useful both for detecting infringement of rights and for avoiding infringement of third party rights. With regard to the issue of detecting infringement, the intellectual property owner is charged with a certain level of vigilance and long delays in detecting an infringement may result in an infringer having an equitable defense against the assertion of the infringed intellectual property right. Thus, you cannot sleep on your rights to the detriment of another. Conversely, one cannot count on ignorance of another's rights as a complete defense to a charge of infringement. As a general rule, one must act in a reasonably prudent manner with respect to his own rights and the rights of another. Frequently, a sales force or distribution network is an excellent source of competitive information. These individuals continuously interface with customers and interact in the marketplace, and are generally aware of competitors' products and services. While the law does not require the engagement of an investigator to search out all possible infringements, it does require an increased level of vigilance once there is a reason to believe there is an infringement.

4. Finally, enforcement is the end result of properly conducting the above steps and identifying an infringer. Enforcement is a process that should not be taken lightly, as it can have consequences for rights being asserted and consequences for the business entity itself. Once an infringer is discovered, there are steps that need to be taken for the purposes of evaluating the infringement and the impact on the business of the intellectual property owner. In other words, knowledge of an infringement carries with it the requirement of action or at least an informed decision of not taking action. The steps needed for the purposes of evaluating the infringement and the impact on the business of the intellectual property owner should be followed in either case.

10.2 Evaluating a Controversy Prior to Commencing Litigation

A number of inquiries should be made prior to commencement of filing suit to enforce intellectual property rights. First, an initial assessment should be made as to whether the intellectual property right can be enforced. This involves identifying the right and checking its validity and enforceability. This evaluation needs to be made even when all steps have been taken to obtain valid enforceable rights, in order to ensure that the right has maintained its enforceability. For example, a patent may have lapsed for failure to pay maintenance fees and will no longer be enforceable under this scenario. Likewise, a trademark that has become abandoned (i.e., the owner is no longer using the mark and has no intention of resuming use) is no longer enforceable. In cases where an intellectual property right has been lost or become unenforceable, it can subsequently be regained or made enforceable again. For example, a patent that has lapsed for failure to pay maintenance fees may be reinstated up to 2 years after the lapse if it can be shown that the delay in payment was unintentional. With any intellectual property right, a check should be made to ensure that nothing has occurred to cause the right to cease being enforceable, and if it is found that such an event has occurred, a further check should be made to find if any steps can be taken to reinstate the right.

All elements of a legal cause of action must be present for successful enforcement. This evaluation is different for each type of intellectual property right. For a patent, it involves asking whether an allegedly infringing device meets each and every limitation of one or more of the patent's claims. For a trademark, it involves asking whether there is a likelihood of confusion between the asserted trademark and the allegedly infringing mark. For a copyright, it involves asking whether there was an actual copying of the work or a portion of the work and whether there is any evidence that the infringer has access to the copyrighted work. With any cause of action, the first step is a check of whether all the elements of the potential claim can be satisfied.

It is always necessary to make a risk benefit analysis of the economic impact of a potential litigation. Litigation can be expensive, and the potential cost of a lawsuit is often difficult to predict. The American Intellectual Property Law Association recently published a report which found that the legal fees in patent litigation could be $600,000 where one million dollars was at stake, and $2.5 million where more than one million dollars was at stake.[1] The estimated litigation cost should be weighed against the damages

---

[1] American Intellectual Property Law Association, *AIPLA Report of the Economic Survey 2007*, American Intellectual Property Law Association (2007).

sought, and the probability of actually obtaining such an amount. It should be kept in mind that in many cases, remedies other than monetary damages may be awarded. A valuation of all potential remedies should be made. For example, if an injunction to stop an infringement is the main relief sought, an analysis of the value imparted to the intellectual property owner by the cessation of the infringing behavior needs to be made before any action is taken. That analysis should include the non-monetary costs of the time and energy that litigation takes away from the normal business operations.

10.3 Remedies and How to Achieve Them

10.3.1 Injunctions

Injunctions are a remedy in most intellectual property lawsuits, especially patent and trademark lawsuits. Injunctions can be sought in addition to monetary damages. As discussed in more detail below, an injunction can be permanent or preliminary.

A party seeking permanent injunctive relief must demonstrate by competent evidence that:

(1) it has suffered an irreparable injury;
(2) other remedies available at law, such as monetary damages, are inadequate to compensate for the injury;
(3) the balance of hardships between the plaintiff and defendant warrant the granting of an equitable remedy; and
(4) the injunction will serve the public interest.

Although permanent injunctions are more common in intellectual property lawsuits because monetary damage amounts are often difficult to ascertain, they are not considered to be automatic upon a finding of infringement. Injunctive relief in an intellectual property matter must meet the same four elements noted above.

10.3.2 Payment of Royalties

One possible remedy for patent infringement is an order that the infringer must pay the intellectual property owner a reasonable royalty. The court may determine what constitutes a reasonable royalty and order payment for past infringement, and, in certain cases involving public interest, order that royalties be paid for the future use of another's intellectual property right.

An order for payment of reasonable royalties may be desirable where the actual damages are difficult to evaluate. Courts will often engage in a weighing of the harms when determining an appropriate remedy in intellectual property cases, and avoid awarding a remedy that will impose an undue burden on the infringer, particularly where the infringement was innocent. In some cases, courts will give an innocent infringer an option between an injunction and an order to pay reasonable royalties until such time as the infringer phases out the use of the owner's intellectual property without suffering excessive damage.

10.3.3 Monetary Damages

Monetary damages can be difficult and expensive to prove in intellectual property litigation because it is often difficult to ascertain the actual loss attributable to and incurred as a result of the infringing behavior. An investigation should be made to determine what damages were incurred, and if they are recoverable. For example, in a suit for trademark infringement, the trademark owner may seek damages for loss of goodwill resulting from the infringement, but measuring those damages can be difficult and involve surveys and experts that may push the costs beyond what is recoverable.

In certain cases, such as infringement of a registered copyright, the owner of the copyright may be entitled to statutory damages without the requirement of proving actual damages if the statutory requirements are met. The copyright statute provides, upon election by the copyright owner, that the court may award anywhere from $750 to $30,000 for each work infringed upon and may further increase the award per infringement up to $150,000 for willful infringement.

10.4 Settling Controversy Without Litigation

10.4.1 Arbitration

Arbitration is a non-judicial form of resolution where the parties select an arbitrator or panel of arbitrators who hear evidence and decide the matter. The proceeding is similar to, but not as formal, as a court trial. The parties can agree that an arbitration award can be binding and enforceable in court and appeals may be permitted.

Arbitration offers various advantages over litigation, and several particular advantages unique to resolution of intellectual property disputes. Arbitration presents the opportunity to have a dispute settled in a more expedient manner than conventional courtroom litigation. Litigation of cases involving intellectual property issues can be long, tedious, and complex. Arbitration may shorten the dispute process, and, because of the shortened time frame and limited need for discovery, reduce the costs compared to those typically associated litigation.

One particular advantage of arbitration with respect to settlement of patent disputes lies in the fact that the arbitrator is chosen by the parties, permitting selection of an arbitrator having a technical background that will facilitate understanding of the patented subject matter. In 1983, the United States Patent Act was amended to provide for the voluntary settlement of patent disputes by binding arbitration.

10.4.2 Mediation

Mediation can be an effective means of settling a dispute more quickly and at a lower cost than litigation, but it is very different from arbitration. The mediator functions differently than a judge, jury, or arbitrator by working with the parties to assist in a negotiated agreement that satisfies the interests of all without any determination of who is right or wrong.

Settlement of disputes by mediation offers several practical advantages. Litigation costs, such as discovery and motion practice, can be greatly reduced or eliminated. Another significant advantage is the abbreviated time lapse between commencement and resolution of the dispute, particularly due to the elimination of appeals which can draw out the litigation process. This is of particular importance in the context of patent litigation, where the right could expire or the technology involved could potentially become obsolete before the dispute is resolved. In addition, parties using mediation for dispute resolution avoid the risk of a complete loss on all counts, and, hopefully, negotiate a resolution that favors the continuation of the business interests of the party.

Mediation can be particularly valuable in resolving disputes over intellectual property rights because it offers the parties an opportunity to come to resolutions that may be unlikely to be obtained in court. Agreements may be reached that allow all parties to exercise an intellectual property right with minimal intrusion on the rights of the other parties. For example, in a dispute over trademark rights, the parties may come to an agreement where each agrees to keep limited use of the trademark, such as by confinement to a particular geographic boundary.

10.5 Litigation

In cases where there is no other solution, litigation may be the necessary means to resolve the dispute.

10.5.1 Selecting a Jurisdiction

Intellectual property lawsuits may be brought in either the state or federal courts. The court used can be dependent on the issue and the authority for the asserted property right. Since trade secret rights most typically arise under state law, those suits are most common in state courts. Patent and copyright disputes are both the subject matter of federal statutes which invoke a federal court jurisdiction. Trademark rights may arise under state or federal law and the authority for such rights will determine whether the suit should be brought in state or federal court. Although most cases involving patent and copyright issues will be brought in federal court, there are exceptions where an intellectual property issue arises collaterally with a state law matter such as interpretation of a license or contract.

In addition to determining whether state or federal court is the appropriate place to bring suit, the defendant must reside in or do business within the state where the suit is brought.

10.5.2 Causes of Action

10.5.2.1 Infringement

Infringement actions are the most common type of litigation involving patents, trademarks, and copyrights. An infringement suit may be brought for either direct or contributory infringement. In a lawsuit brought for direct infringement the party allegedly engaging in the infringing activity is named as the defendant.

For patents, infringement results from making, using, selling, offering to sell, or importing into the United States an invention within the scope of the patent claims without the patent owner's permission. For trademarks, infringement results from the unauthorized use of the same mark or a mark that has a likelihood of creating confusion between that mark and the plaintiff's mark. For copyrights, infringement results when the infringer had notice of or access to the copyright work and engaged in actual copying of all or part of the copyrighted work.

In some circumstances, the infringement may result from activity that is not directly infringing, but it encourages or enables infringement by a third party. In these cases, the intellectual property owner may bring a lawsuit against a party for inducing or contributory infringement (i.e., indirect infringement). Such causes of action for indirect infringement can be based upon various activities. For instance, the defendant may be making or selling product that is intended for an infringing use or is sold with instructions that provide directions for another to infringe. This situation is common in both the patent and copyright contexts. In either situation it may be undesirable to file suit against the direct infringers as there may be a large number of defendants or they may be customers, and the direct infringers are typically less likely to have recoverable funds than the indirect infringer.

Indirect infringement frequently arises where a patent contains only method claims. Someone making or selling a product that does not itself infringe the patent claims may encourage its use in a way that does infringe the claimed method. The patent owner may not be able to sue the maker or seller for direct infringement of the method, but could sue the maker or seller for indirect infringement if those activities induced or contributed to others to infringe a patent.

10.5.2.2 Declaratory Judgments

A party having a real interest to capitalize on subject matter that is alleged to be covered by another party's intellectual property right may file suit seeking a declaration that there is no infringement of the intellectual property right in question, the right in question is invalid, or that the right in question is not enforceable.

The decision to seek a declaratory judgment requires serious thought. The owner of the rights in question may not pursue the potential infringement based on its own policy reasons or may no longer have a sufficient commercial interest in them to expend the costs to pursue litigation. Thus, any decision to file a declaratory judgment action must under go the same analysis applied to any other litigation.

10.5.2.3 Trademark Dilution

Trademark dilution originated as a state law cause of action pertaining to common law trademark rights. In 1995, it was codified in the Federal Trademark Dilution Act and made applicable to federally registered trademarks. Dilution actions can only be brought by owners of famous marks. To make out a cause of action, the plaintiff must show that the defendant adopted the use of a mark that tends to dilute the distinctive quality of plaintiff's famous mark. If such a showing is made, the plaintiff can obtain an injunction against the defendant's use of the mark, even if a showing of likelihood of confusion is absent.

10.5.2.4 Misappropriation of Trade Secrets

To bring an action for misappropriation of trade secrets, a plaintiff must show the existence of a trade secret, and that the defendant misappropriated the trade secret. Under the Uniform Trade Secrets Act (UTSA), which many states follow, a trade secret is any information having independent economic value due to its nature and the fact that it is not commonly known, for which reasonable efforts are made to maintain secrecy. Misappropriation occurs where the defendant either acquired the information by improper means, or received the information with knowledge that it was derived by improper means or disclosed in breach of a duty to keep secret. Trade secret law does not take the "strict liability" approach of patent and trademark laws, and requires some wrongful behavior on the defendant's part in order to impose liability. It should also be kept in mind that a party who "reverse engineers" the trade secret cannot be held liable for misappropriation of trade secrets, as this is not considered an "improper means".

10.5.3 Preliminary Injunctions

Seeking a preliminary injunction at the outset of the litigation may be desirable to prevent further infringement during the proceedings. Such an injunction is temporary, and is typically only awarded upon a showing that the plaintiff has a strong likelihood of success on the merits, along with the potential to suffer irreparable injury if the allegedly infringing activity is permitted to continue during the proceedings. This can be a difficult burden for a plaintiff, particularly in intellectual property lawsuits where the court is often charged with detailed factual evaluations or subjective multifactor tests. Nonetheless, a case can be presented that warrants a preliminary injunction in order to mitigate damages.

10.5.4 Discovery

Discovery is a highly important phase of intellectual property litigation and its focus will vary depending on what type of intellectual property right is at issue. The Federal Rules of Civil Procedure govern intellectual property proceedings in federal court, while state rules will govern those cases that are brought in state courts.

Depositions, interrogatories, requests for admissions, and document requests are all available as discovery tools in intellectual property litigation. The Federal Rules of Civil Procedure provide for discovery of "any non-privileged matter that is relevant to any party's claim or defense".[2]

10.5.5 Summary Judgment

Motions for summary judgment can be an efficient means for bringing litigation to an early conclusion. Federal Rule of Civil Procedure 56(c) states that a party is entitled to summary judgment where "the pleadings, the discovery and disclosure materials on file, and any affidavits show that there is no genuine issue as to any material fact and that the movant is entitled to judgment as a matter of law."

Having a motion for summary judgment granted can be challenging in intellectual property cases where a large number of complex issues are often present and the evidence may be subject to different interpretations. One example is the case in which a defendant moves for summary judgment holding that the plaintiff's intellectual property right is invalid or unenforceable. While there are some instances where the controlling law is so clear that there is no factual dispute, these issues frequently involve the need for testimony and an evaluation of credibility by a fact finder. Although a grant of summary judgment may be difficult at times, it may be desirable as a means of framing the issues in an effort to expedite the proceedings.

10.5.6 Trial

The trial phase of litigation can be long and expensive in both time and money. The more issues and the more complex the issues, the longer the trial is likely to take. Patent and trade secret cases can have lengthy trials due to the need to educate the fact finder on the technology at issue. Trademark and copyright trials can also be time consuming; however, some of these cases can involve simple issues that are easily comprehended by the fact finder.

Experts are especially important in intellectual property litigation, particularly in trade secret and patent litigation where complicated technological issues are likely to be present. An expert witness in a patent case may be needed to testify as a "person having ordinary skill in the art". An expert can also help by assisting the judge and jury in gaining an understanding

---

[2]Federal Rule of Civil Procedure 26(b).

of technology involved in the case. Experts are also frequently used in the economic area to establish or refute damages.

10.5.7 Costs

Some statutes make attorney's fees and costs available as remedies to intellectual property owners involved in litigation. Both the Patent Act and Trademark Act make attorneys' fees available in "exceptional cases". The Copyright Act permits a court in its discretion to award litigation costs including attorney's fees to the prevailing party. Courts usually interpret each of the provisions as applying to cases where willful infringement has been found. US courts do not routinely grant fee shifting awards and litigation should not be viewed as a "loser pays" situation.

10.6 Proceedings in the US Patent and Trademark Office

10.6.1 Trademark Oppositions

A trademark opposition is an interparties proceeding that may be requested by any party who believes there is a potential to be damaged by the registration of a trademark. After a trademark application is examined and found allowable, the US Patent and Trademark Office publishes the mark in its *Official Gazette* to put the public on notice of the potential registration of the mark. A potential opposer has 30 days from the date of publication to file an opposition, or request an extension of time for filing an opposition. Typical grounds for filing an opposition include likelihood of confusion with the opposer's mark, genericness or descriptiveness, abandonment, and fraud.

After an opposition is filed, the trademark applicant/owner responds by filing an answer asserting any defenses. Frequently, the answer to an opposition will assert a counterclaim to cancel a registered mark of the opposer which forms the basis of the opposition. The opposition proceeding is held before the Trademark Trial and Appeal Board. This is an administrative panel, rather than a judicial panel, with expertise in trademark matters. The proceeding is conducted similar to a federal court proceeding, but is less formal. The PTO proceeding concludes with a Board decision either refusing the registration or permitting the pending registration to go forward to registration. The losing party may appeal the Board's decision to the US Court of Appeals for the Federal Circuit. The denial of a registration does not mean that the mark cannot be used. The right to use the mark requires a judicial determination; however, the results of an opposition can often lead to a settlement between the parties.

10.6.2 Patent Interference

A patent interference proceeding is a highly technical interparties proceeding that is declared when two inventive entities are claiming the same patentable invention. In order to have interfering subject matter, there must be a reasonable possibility that the first entity to file a patent application was not the first to invent. This is a condition of establishing a right of priority to the inventive claims.

Priority of invention is decided in an interference conducted before the Board of Patent Appeals and Interferences, another administrative panel within the USPTO. The party who filed the earlier patent application is designated as the "senior party," and the later filer is known as the "junior party." For determining priority, the concept of invention is broken up into two parts: (1) conception of the claimed invention and (2) due diligence in the reduction to practice of the claimed invention. Conception is the mental element of invention. Reduction to practice happens subsequent to conception and it may be based on an actual reduction to practice or constructive reduction to practice. For an actual reduction to practice, the inventor must construct a working embodiment of the invention. Constructive reduction to practice results from the filing of a patent application.

In determining priority of invention, the junior party has the burden of proving an earlier date of invent. This can be done in two ways. First, the junior party can show an earlier date of conception coupled with diligent reduction to practice, either actually or constructively. The senior party's evidence may establish a conception date prior to the patent application date and a subsequent reduction to practice.

At the conclusion of the interference proceeding, the Board issues a decision awarding one party priority to the subject claims and the applications are returned to the examiner, who are bound by the Board's findings, for final examination.

10.6.3 Patent Reexamination

A reexamination may be ordered at any time after a patent has been granted. Anyone, including the patent owner, may file a request for reexamination which shows the existence of a substantial new question of patentability. Reexaminations are most frequently requested in two situations. First, a competitor or potential infringer seeks to have a patent invalidated and removed from the commercial arena. Second, the patent owner wants to enforce the patent, but has concerns about the patent's validity. Reexamination may resolve that issue, and it has the potential to strengthen the patent by bringing the additional prior art not applied during examination to the attention of the US Patent and Trademark Office. Reexamination proceedings may be conducted either as an ex parte proceeding or as an inter partes proceeding.

Chapter 11
Licensing Intellectual Property Rights

Licensing intellectual property can be used to learn and use other's technology, allow others to access technology, and create a revenue stream. For example, a food technology company may need to license proprietary manufacturing equipment from others. Conversely, a company may consider licensing its product along with a name and marketing campaign to a third party to create a larger food distribution network and generate revenue from a larger market. Finally, IP rights can be licensed or, in some cases, cross-licensed to resolve litigation.

11.1 What Is an Intellectual Property License?

An intellectual property (IP) license is a contract between two parties allowing the licensee to use at least a portion of the licensor's intellectual property in exchange for some consideration. Some examples of what the licensor receives are: a lump sum, multiple payments, royalty stream, goods, services, a cross-license to the licensee's IP, or combinations thereof. License payments may be tied to the sale of goods or services provided or sold (e.g., as a percentage of gross or net revenue for a product that the licensee is selling). Technology transfer agreements, which are often used by universities to capitalize on research and development, often involve an up front minimum payment as well as a stream of royalty payments.

> *Example 1*: Tastewell requires specialized equipment available only from ACME manufacturing (who has patented the equipment) in order to produce its new fruit and cheese product.
>
> Tastewell will need to obtain a license to use the equipment from ACME manufacturing. If the equipment is purchased from ACME, it should include a license to use the equipment for its intended purpose in the purchase price. However, equipment manufacturers of specialized

R.W. O'Donnell et al., *Intellectual Property in the Food Technology Industry*, 113
DOI: 10.1007/978-0-387-77389-6_11, © Springer Science+Business Media, LLC 2008

packaging equipment have recently been granting only limited licenses for equipment installed for a particular use and adding a "per item processed" royalty on top of the purchase price. Benefits in such arrangements may include a lower initial capital expense and the right to improvements. The result is something akin to a specialized equipment lease/purchase agreement that can have benefits to both parties.

Example 2:   Tastewell decides the equipment and license from ACME are too expensive and that it can modify some generic equipment to meets its needs.

An issue that often arises is whether the modifications will result in the generic equipment becoming infringing. This may still require a license from ACME or a third party, and an infringement clearance search should be considered.

Example 3:   After Tastewell has installed equipment from ACME and used it for several years, it wants to modify the equipment to fill different types of packages. ACME has separate patents to cover the modifications that Tastewell wishes to incorporate.

Tastewell will need to obtain a license to use the equipment modifications, even though the original equipment and its use were licensed.

11.2 Factors to Consider in an IP License

There are numerous factors that need to be considered in an IP license which are unique depending on the facts and circumstances presented. These may include what rights to license, territory considerations, whether the license should be exclusive, sole or non-exclusive, the amount and structure of the royalty, indemnification from damage caused by the other party, etc. Special consideration is also required if the license relates to the use of a licensor's trademark. These issues are discussed briefly below:

11.2.1 Identification of Rights to be Licensed

Patent rights include the right to make, use, sell, or offer for sell a patented article or process. These rights are territorial, and are limited to the country in which the patent is granted. Trademark rights are the right to use a trademark in connection with goods or services, and are also territorial by country. Other rights can provide access to a proprietary or trade secret technology or know-how.

Example: Tastewell wants BigFoods Distribution to distribute its new cheese product, and use the name Tropical Island Cheese™.

Tastewell should identify the rights it is going to license, which could be one or more of patents or pending patent applications directed to the Tropical Island Cheese product, technology and know-how related to manufacture of the product, trademarks associated with the product, whether registered or unregistered, and possibly trade secrets, such as the product formula.[1] As patent rights are for a limited term (the maximum life of a patent being 20 years from the earliest filing date) and any patent license would automatically terminate once the patent expires, it is beneficial to also license intellectual property rights that do not expire, such as a trademark and/or technology and know-how for manufacturing the product so that the license does not have a fixed term.

11.2.2 Restrictions

A license allows the licensor to restrict a licensee's activities to something less than an unlimited right to use the licensed intellectual property. For example, a patent licensor may choose to grant a licensee one or more of its rights to make, use, sell, or offer for sale a patented product.

---

[1] Divulging trade secret information, even under the terms and safeguards noted in a license can involve the risk of loss of the trade secret, and often it is beneficial to structure the license so that a trade secret is not divulged. One possible solution is to structure the license so that the licensor provides a trade secret ingredient package under the terms of the license for use in making the Tropical Island Cheese product. The most famous example of this is the Coca-Cola® syrup that is delivered to Coca-Cola® bottlers who have licensed the right to produce Coca-Cola® soda, but who never gain access to the trade secret.

Another restriction is whether the licensor is granting an exclusive, sole, or non-exclusive license. An exclusive license is similar to an assignment of the IP rights to the licensee, and the licensor foregoes the ability to use its own IP rights in favor of the licensee. This type of license has the most value to the licensee because it prevents all competition from using the licensed IP rights, and is often granted by research institutions that have no intention of commercially exploiting an invention. A sole license, in contrast to an exclusive license, allows the licensor to continue to use the IP rights, but limits the licensor from granting any further licenses to third parties. A non-exclusive license allows the licensor to grant multiple licenses to third parties.

Example: Tastewell wants to distribute its Tropical Island Cheese product itself, and also license BigFoods to distribute the product.

Tastewell should consider giving a sole or non-exclusive license to BigFoods, and may wish to consider other types of restrictions, as noted above, to protect a certain portion of the market.

Territorial restrictions are another common restriction in license agreements. A license should explicitly define the licensed territory. For example, a company may be able to service a limited geographic territory and would not want to grant any rights to any third party to compete in this same area.

Field of use limitations are also frequent restrictions included in manufacturing equipment licenses, and may limit the use of the equipment or ingredients to certain product types, such as food or non-food, package size as well as other types of limitations. This allows a licensor to grant multiple licenses to exploit the licensor's IP more effectively.

11.2.3 Consideration

Consideration between parties to a contract regarding intellectual property rights can take many forms, but usually involves some form of payment. The most common are a lump sum payment, a stream of royalty payments (most commonly based on the number of products sold), as well as a combination of both. A lump sum payment can be used in a number of circumstances, such as settlement of past infringement, or when the technology is being transferred and the licensor is going to be working in the field. This type of royalty shifts the entire risk of success onto the licensee since the licensor receives payment no matter what happens. As a result, an upfront lump sum payment is often discounted from the amount that could be obtained by taking a stream of royalty payments over time based on sales volume of the product. A stream of royalty payments over time based on sales volume miti-gates the risk that the licensee must take, because the licensee does not make any payments unless the product is selling. The licensor stands to receive a greater stream of royalty revenue if the product does well. A combination of some upfront lump sum payment to offset research and development costs incurred by the licensor along with a reduced stream of royalty payments is sometimes used to strike a balance for a licensor that wants some immediate payment, and also wants to share in the success of the product.

Many licenses are structured so that a minimum royalty payment must be made in a given time period in order to maintain the license in force. This prevents a licensor from obtaining, for example, an exclusive license and then failing to take action to manufacture, advertise, or sell the product. Failure to meet minimum payments can act as a trigger for automatically changing a license from exclusive to non-exclusive, or for terminating the license.

Setting royalty rates is considered to be an art more than a science. While some reference materials are available that provide "typical ranges" of roy-alty rates for a particular field, each situation must be reviewed independently based on all of the available information, such as the predicted market size, ramp-up time, whether production is capital intensive, etc. Additionally, the relative size and strength of the parties is typically taken into account.

In addition to royalties, other non-monetary items can form the consideration. Cross-licenses may be obtained if both parties possess intellectual property rights that the other party desires. This may replace or offset some or all of the monetary consideration.

11.2.4 Maintenance of IP Rights

There are a number of terms that should be included in a license agreement that relate to the maintenance of the licensed intellectual property rights. For patent rights, this should include the requirement that the licensee include the appropriate patent marking on any goods sold that are covered by the patent. Lack of patent marking can limit claims for damages against infringers. Depending on whether the license granted is an exclusive license, the licensor may also require the licensee to pay patent maintenance fees. In some situations, a licensee may even take over prosecution of pending patent applications, which may benefit the licensee especially in the situation where the licensor is an individual or has limited resources to prosecute the patent application.

In trademark licenses, in order for the licensor to maintain its trademark rights, the agreement should require that any goods and services using the trademark identify that it is being used under license from the licensor. Additionally, to maintain the mark, the trademark license should have provisions for inspecting any products which use the trademark to ensure that the quality of the goods is consistent with the licensor's standards. If a licensee does not maintain quality control over the usage of the trademark, not only can the licensor suffer damage due to poor quality goods being associated with the mark, but the trademark can also be lost.

11.2.5 Other Terms

License agreements will generally include a number of other terms, some of which are discussed below.

(a) Representations and warranties are typically included from the licensor to the licensee, and should include a statement that the licensor is in fact the owner of the IP rights being licensed and has the right to grant the license. This offers some protection to the licensee from fraudulent transactions. The licensor should also warrant that they believe the IP rights to be valid, and should identify any known challenges to the IP rights. While a licensee should be conducting its own due diligence review of the IP rights being licensed, if the licensor fails to reveal this type of information, it can provide grounds for rescinding the license if the IP rights ultimately prove to be invalid.

(b) A license will often include terms regarding ownership and/or cross-licensing of further developments. From the licensor's perspective, this can be important if the licensor is also producing a product under the IP rights, and wants to have the benefit of any of the licensee's improvements, which are based on the licensor's underlying IP rights. From the licensee's perspective, having rights to improvements may eliminate the need for a further license relating to the same products or services being provided under the licensed IP rights.

(c) The burden for obtaining regulatory approval, for example, from the FDA, should be designated in the license. For new product types, this can be a time-consuming process, and the burden for obtaining the required approvals should be specified. If the burden is placed on the licensee, this can be used as a negotiating point to obtain a reduced royalty, at least during the time it takes to obtain the approval.

(d) In the event that the parties ultimately disagree over the meaning or enforcement of any provision in the license agreement, some form of dispute resolution should be included in the agreement. This should not only include a choice of law, but a forum for any action or arbitration that will take place. In order to avoid the time and cost of a court proceeding, many agreements now call for binding arbitration of any dispute between the parties. A "loser pays" provision has also become standard in most license agreements to avoid meritless claims.

(e) A termination clause is also standard in any license agreement, and should, in addition to setting any fixed or renewable term limits for the license, include a list of circumstances or actions that will result in automatic termination of the license. Automatic termination will generally occur for non-payment of any royalties due, failure to launch or market product within a predetermined time limit, or bankruptcy of the licensee. Termination may also occur for breach of any other terms of the license agreement, generally after a notice and cure period.

A sample license agreement between Tastewell and BigFoods which includes many of the above items is attached in Appendix E.

Appendix A

CONFIDENTIAL DISCLOSURE AGREEMENT

This Agreement made by and between Tastewell Industries ("TASTE-WELL") having a principal place of business at _____, and BigFoods, Inc. ("BIGFOODS"), having a principal place of business at _____.

WITNESSETH THAT:

WHEREAS, TASTEWELL has developed a new cheese product, (hereinafter referred to as "the invention"), and is in possession of certain related confidential and proprietary information (hereinafter referred to as "proprietary information");

WHEREAS, TASTEWELL is interested in disclosing the invention and proprietary information to BIGFOODS, in confidence, for further evaluation for product production and distribution rights; and

WHEREAS, BIGFOODS is interested in receiving such information, in confidence, for conducting such further evaluation.

NOW, THEREFORE, for and in consideration of the foregoing premises, and of the mutual promises set forth below and for other good and valuable consideration, the receipt and sufficiency of which is hereby acknowledged, the parties, intending to be legally bound, hereby agree as follow:

1. TASTEWELL agrees to disclose to BIGFOODS in confidence proprietary information relating to the invention.
2. BIGFOODS agrees to accept and hold in confidence any and all proprietary information disclosed by TASTEWELL under this Agreement, except:

    (a) information which at the time of disclosure can be shown to have been in the general public knowledge;

(b) information which, after disclosure, becomes part of the public knowledge by publication or otherwise, except through breach of this Agreement by BIGFOODS;

(c) information which BIGFOODS can establish by competent proof was in its possession at the time of disclosure by TASTEWELL and was not acquired, directly or indirectly, from TASTEWELL; and

(d) information which BIGFOODS receives without restriction from a third party, provided that such information was not obtained by said third party, directly or indirectly from TASTEWELL.

3. BIGFOODS agrees that the proprietary information received from TASTEWELL shall not be used by BIGFOODS, other than for evaluation and consideration for use as noted above and as otherwise agreed between TASTEWELL and BIGFOODS.

4. This Agreement shall not be construed as granting any license or any other rights to BIGFOODS.

5. BIGFOODS agrees to restrict access to the proprietary information to those employees, agents and representatives who are engaged in that actual evaluation of the invention on a need-to-know basis and will require all such employees, agents and representatives to agree to maintain the proprietary information in confidence.

6. BIGFOODS agrees that any improvements or modifications developed by BIGFOODS in connection with the invention shall belong to TASTEWELL, and BIGFOODS shall assign all rights in any such improvements or modifications to TASTEWELL.

7. Upon completion of its evaluation, BIGFOODS agrees to return to TASTEWELL all information concerning the invention, including all photographs, diagrams, drawings, descriptions, prototypes, and notes, and any copies thereof.

This Agreement shall be binding upon and shall inure to the benefit of and be enforceable by and against the respective heirs, legal representatives, successors, assigns, subsidiaries, and affiliated or controlled companies of the parties hereto.

This Agreement shall be construed, interpreted and applied in accordance with the law of the State of _____. With respect to the subject matter of this Confidential Disclosure Agreement, the foregoing constitutes the entire and only understanding between the parties, and this Confidential Disclosure Agreement supersedes any prior or collateral agreements or understandings between the parties with respect to confidentiality.

IN WITNESS WHEREOF, the parties hereto have caused this Agreement to be executed as of the last date written below.

BigFoods, Inc.

Date: _____   By: _____
                                 Name:
                                 Title:

Tastewell Industries

Date: _____   By: _____
                                 Name:
                                 Title:

Appendix B

ASSIGNMENT

Dr. Richard Curd, residing at _____, a citizen of the United States (hereafter the undersigned), is the inventor of _____ for which the undersigned executed an application for United States Letters Patent, U.S. Patent Application No. _____, filed _____, 200___.

The undersigned hereby authorizes assignee or assignee's representative to insert the Application Number and the filing date of this application if they are unknown at the time of execution of this assignment.

Tastewell Industries, a Delaware Corporation, having a place of business at _____ , (hereafter referred to as the assignee), is desirous of acquiring the entire right, title and interest in said invention, all applications for and all letters patent issued on said invention.

For good and valuable consideration, the receipt and sufficiency of which is acknowledged, the undersigned, intending to be legally bound, does hereby sell, assign and transfer to the assignee and assignee's successors, assigns and legal representatives the entire right, title and interest in said invention and all patent applications thereon, including, but not limited to, the application for United States Letters Patent entitled as above, and all divisions and continuations thereof, and in all letters patent, including all reissues and reexaminations thereof, throughout the world, including the right to claim priority under the Paris Convention or other treaty.

It is agreed that the undersigned shall be legally bound, upon request of the assignee, to supply all information and evidence relating to the making and practice of said invention, to testify in any legal proceeding relating thereto, to execute all instruments proper to patent the invention throughout the world for the benefit of the assignee, and to execute all instruments proper to carry out the intent of this instrument.

The undersigned warrants that the rights and property herein conveyed are free and clear of any encumbrance.

EXECUTED under seal on this _____ day of _____, 200_

at

_____.

(Place)

Witness:

_____   _____(L.S.)
                                                   Dr. Richard Curd

State of

                              ss.

County of

On this_____ day of _____, 200__ before me personally appeared Dr. Richard Curd, to me known to be the person described herein and who executed the foregoing instrument, and acknowledged that he executed the same knowingly and willingly and for the purposes therein contained.

Witness my hand and Notarial seal the day and year immediately above written.

                                        _____
                                        Notary Public

My Commission Expires:

Appendix C

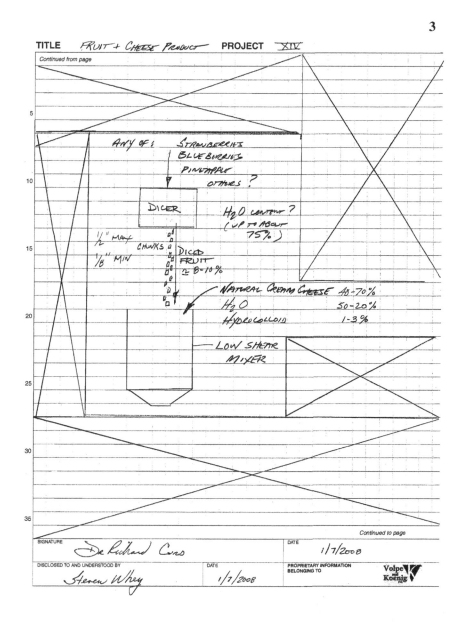

TITLE FRUIT + CHEESE PRODUCT   **PROJECT** XIV

Continued from page

ANY OF: STRAWBERRIES
         BLUEBERRIES
         PINEAPPLE
         OTHERS ?

DICER

H_2O CONTENT ?
(UP TO ABOUT
75%)

½" MAX
⅛" MIN   CHUNKS

DICED
FRUIT
≈ 8-10%

NATURAL CREAM CHEESE 40-70%
H_2O 50-20%
HYDROCOLLOID   1-3%

LOW SHEAR
MIXER

Continued to page

SIGNATURE Dr Richard Curd   DATE 1/7/2008

DISCLOSED TO AND UNDERSTOOD BY   DATE 1/7/2008

Steven Whey

PROPRIETARY INFORMATION
BELONGING TO   Volpe and Koenig

127

Appendix D

---

(1) Title of Invention:

(2) Description of the Invention:

(3) First written description of the invention was made by: Date:

(4) First drawing of the invention was made by: Drawing No.:   Date:

(5) First model was made by: Date:

(6) Invention was first disclosed to: Date:

(7) First effort to sell the invention: Date:

(8) First use of invention: Date:

(9) Description of circumstances(s) and date(s) of any disclosure(s) that are not described above:

---

8 VOLPE and KOENIG, P.C. 1995

(10)   If you know of other products, processes or machines like yours, describe them and how your invention differs from them.

(11)   Identify any known patents and/or other publications which relates to your invention. (Attach copies, if available.)

(12)   In the case of a process invention has a product ever been made by the new process? Yes__ No__. If yes, has the product ever been offered for sale, sold, or used? Yes__ No__. If yes, please describe the circumstances with dates.

(13)   If anyone other than a named inventor had anything to do with this invention, state the person(s) full name, company affiliation, address and title and describe that involvement.

(14)   If this invention is assigned or is to be assigned, identify the assignee.

Inventor:_____   Inventor:_____
         (Print full name)   (Print full name)

Citizenship:_____   Citizenship:_____

Home Address:_____   Home Address:_____

              _____   _____

              _____   _____

Telephone No.:(   )   -   Telephone No.: (   )   -

Social Security No.:_____   Social Security No.:_____

Signature_____Date_____   Signature_____Date_____

Attachments to this form:

Appendix E

(SAMPLE) LICENSE AGREEMENT

This Agreement made this ___ day of _____, 2008 by and between Tastewell Industries, a Pennsylvania corporation, having a place of business at _____ (hereinafter "TASTEWELL") and BigFoods, Inc., a Delaware corporation, having a place of business at _____ (hereinafter "BIGFOODS").

WHEREAS, TASTEWELL has developed a product (hereinafter "the TASTEWELL Product") intended for distribution and sale in a nationwide market and has expertise in producing such product; and

WHEREAS, BIGFOODS has a nationwide distribution network and is a manufacturer and distributor of similar types of products and is desirous of obtaining the TASTEWELL Product and know-how as well as the right to manufacture, distribute and sell the TASTEWELL Product;

NOW, THEREFORE, in consideration of the above and of the mutual covenants and obligations contained herein, and intending to be legally bound, the parties hereto agree as follows:

1. DEFINITIONS

1.1 "Licensed Goods" shall refer to TASTEWELL's Tropical Island Cheese product as described in TASTEWELL's U.S. Patent Application No. XX/XXX,XXX.

1.2 "Licensed Territory" shall mean North and South America.

1.3 "Affiliate", with respect to either party, shall mean any person or entity who controls, is controlled by, or is under common control with a party to this agreement, and includes, but is not limited to, parent corporations, subsidiaries, and sister corporations.

1.4 "Net Selling Price" is the gross selling price (i.e., the dollar amount) actually paid to and collected by BIGFOODS in a bonafide arms-length transaction consummated or intended to be consummated by transferring title in a Licensed Good, less:

(a) returns actually credited;
(b) actual losses experienced by BIGFOODS as a result of credits issued for such things as expired shelf life; and
(c) shipping charges separately charged to transferee;
      however, no deduction shall be made for any other costs incurred, such as, but not limited to, costs of manufacture, sales, distribution, or exploitation of the Licensed Goods.

2. LICENSE

2.1 TASTEWELL hereby grants BIGFOODS and Affiliates a sole license for the manufacture, distribution and sale of the Licensed Goods in the Territory. TASTEWELL will provide technological support and know-how to manufacture the Licensed Goods to BIGFOODS.

3. ROYALTY

3.1 BIGFOODS agrees to pay a royalty to TASTEWELL of _ percent (_%)on the Net Selling Price of the Licensed Goods. Royalties shall be payable on sales made during the period starting on the date of this agreement and ending on January 1, 2009, and for each annual period following therefrom, during the term or extended term of this agreement.
3.2 Royalty payments made under this agreement shall be made within sixty (60) days after the end of each quarterly period during the term of this agreement.
3.3 BIGFOODS agrees to submit to TASTEWELL within sixty (60) days after the end of each quarterly period a written royalty report setting forth the amount of royalties due for the preceding quarterly period and the manner in which BIGFOODS calculated said royalties. BIGFOODS agrees to keep complete records covering all royalty bearing activities specified in this agreement in sufficient detail under its current accounting to enable the royalties payable hereunder to be determined and verified.

4. AUDITS

4.1 The parties hereby agree that TASTEWELL shall be permitted, at TASTEWELL's expense, to have a mutually agreed upon independent certified public accountant audit each royalty report submitted by BIGFOODS to TASTEWELL, within six (6) months from the date it is received by TASTEWELL. BIGFOODS shall make its records available to said accountant and cooperate by providing all available records essential to the verification of the report being audited and said accountant shall maintain confidential all information learned in the course of examining BIGFOODS' records, with the exception of a report to TASTEWELL with the accountant's findings as directly related to BIGFOODS' obligations to make royalty reports and payments. In the event of a finding by such accountant of a material variance with the report issued by BIGFOODS, then BIGFOODS shall reimburse to TASTEWELL the audit costs paid to the accountant and pay to TASTEWELL any additional royalties determined to be due.

5. PATENTS

5.1 Should BIGFOODS obtain information that any patents owned by TASTEWELL are or may be infringed, it shall provide such information to TASTEWELL, but shall have no further responsibility or obligation. Any patents obtained by BIGFOODS relating to the Licensed Goods or improvements to the Licensed Goods shall be the property of TASTEWELL. BIGFOODS shall promptly review any papers and execute, acknowledge and deliver all such papers as may be necessary or desirable, in the sole discretion of TASTEWELL, to obtain or maintain patent protection for the Licensed Goods and to confirm the ownership of any such patent by TASTEWELL.

6. TRADEMARKS

6.1 BIGFOODS shall use the trademark "TROPICAL ISLAND CHEESE", in such form as specified in writing by TASTEWELL, on the Licensed Goods in the Licensed Territory. BIGFOODS shall also use in connection with the trademark a "^{TM}" or, where U.S. Federal Trademark Registration has been obtained, an "®". Once approved, BIGFOODS shall not depart from the approved form of the "TROPICAL ISLAND CHEESE" mark on any materials requiring approval without the approval of TASTEWELL in accordance with paragraph 6.4 of this Agreement.

6.2 In order to assure the development, manufacture, appearance, quality and distribution of the Licensed Goods is consonant with the quality of the trademark, TASTEWELL retains the right to review the Licensed Goods.

6.3 BIGFOODS shall submit to TASTEWELL for approval samples of all Licensed Goods prior to any distribution or sale thereof by BIGFOODS.

6.4 Any such submission of the Licensed Goods for approval which is not disapproved within fifteen (15) days shall be deemed approved. Any disapproval by TASTEWELL shall be submitted to BIGFOODS in writing within the aforesaid fifteen (15) days together with remedial changes which would remedy such disapproval.

7. ENFORCEMENT

7.1 If TASTEWELL obtains patent protection for the Licensed Goods, and any such patent is infringed by a third party, BIGFOODS and TASTEWELL may take appropriate action to suppress such infringement. As patent owner, TASTEWELL shall have the first right, but not the obligation to take action. In the event that TASTEWELL takes action against an alleged infringer, TASTEWELL shall be entitled to the entire recovery. If BIGFOODS requests TASTEWELL in writing to suppress any infringement, identifying in the request the infringer and the circumstances of the infringement, and TASTEWELL fails to file suit against the identified infringer or to otherwise take action to cause the identified infringement to cease within sixty (60) days of BIGFOODS' request, BIGFOODS shall have the right to file suit against and to negotiate and enter into a settlement with the identified infringer. If BIGFOODS files suit, TASTEWELL is under no obligation to bear any cost of the suit. TASTEWELL, at BIGFOODS' expense, shall join in the suit and render assistance and sign all papers, as may be reasonably required in connection with such enforcement. TASTEWELL shall be entitled to 20% of any court awarded or lump sum recovery, less costs, as a result of any enforcement of such patents by BIGFOODS.

7.2 If BIGFOODS becomes aware of a third party infringement of the "TROPICAL ISLAND CHEESE" mark in connection with food products in the Licensed Territory, BIGFOODS shall provide notice and the details of such infringement to TASTEWELL, and TASTEWELL shall take appropriate action to suppress such infringement.

8. TECHNOLOGICAL SUPPORT, QUALITY CONTROL AND PERFORMANCE ASSURANCE TESTING

8.1 TASTEWELL shall provide instructions and know-how for the production of the Licensed Goods.

8.2 TASTEWELL and BIGFOODS agree to jointly develop a Quality Assurance Plan to meet all applicable agency regulations and certifications for the Licensed Goods. BIGFOODS shall test each production batch of the Licensed Goods and shall maintain test records in accordance with the Quality Assurance Plan.

9. PRODUCT LIABILITY AND WARRANTY CLAIMS

9.1 BIGFOODS shall assume all liability for all claims of any nature with respect to the Licensed Goods distributed or sold by BIGFOODS BIGFOODS hereby agrees to indemnify, defend and hold TASTEWELL harmless, from and against any loss, liability, damages and expenses (including reasonable attorney's fees and expenses) which may be incurred or for which TASTEWELL may be obligated to pay or for which TASTEWELL may become liable or be compelled to pay in any action, claim or proceeding against BIGFOODS and/or TASTEWELL for or by reason of any acts, whether of omission or commission, that may be claimed to be or are actually committed or suffered by BIG-FOODS in connection with BIGFOODS' performance of this agreement. The provisions of this paragraph and the obligations under the same shall survive the expiration of this agreement. BIGFOODS shall maintain and procure at BIGFOODS' expense a comprehensive general liability policy including, but not limited to, contractual advertising and products liability coverage's with a policy limit of not less than $__ million per occurrence. Such policy shall be in full force during the entire term of this agreement and shall be placed with a responsible insurance carrier and shall name TASTEWELL as an additional insured and provide for at least thirty (30) days prior written notice to TASTEWELL of the cancellation or modification of such policy.

9.2 In the event that BIGFOODS does not obtain and maintain the aforesaid policy continuously in effect, upon prior written notification to BIG-FOODS, TASTEWELL may obtain such insurance policy on behalf of BIGFOODS. All premiums for such insurance policy shall be deducted from royalties due under paragraph 2 of this agreement.

10. TERM AND TERMINATION

10.1 This agreement will have an initial term of seven (7) years from the execution date.

10.2 This agreement shall automatically be renewed for subsequent five (5) year terms, subject to the right of either party to terminate upon written notice one (1) year prior to the expiration of the initial term or any subsequent renewal term.

10.3 Notwithstanding the aforesaid, this agreement shall be subject to the rights of earlier termination by the party indicated as hereinafter set forth:

   (a) By TASTEWELL in the event that BIGFOODS fails to make royalty payments following ten (10) days prior written notice and demand to cure from TASTEWELL; provided, however, if BIG-FOODS cures such default within the ten (10) day period then such notice shall be of no force and effect; or

   (b) By either party in the event that the other party breaches any other material obligation imposed upon it under this Agreement and fails to cure such breach within a period of thirty (30) days after notice and demand for cure from the party not in breach; provided,

however, if the defaulting party cures its breach within the thirty
(30) day period, then such notice shall be of no force and effect; or

(c) By either party immediately upon the other party becoming bank-
rupt, insolvent, making an assignment for the benefit of creditors,
applying for or consenting to the appointment of a trustee or re-
ceiver or if bankruptcy proceedings are instituted against BIG-
FOODS.

11. RESOLUTION OF DISPUTES BETWEEN THE PARTIES

11.1 This agreement shall be deemed entered into the Commonwealth of
Pennsylvania and shall be construed and governed solely by the laws
of Pennsylvania.

11.2 In the event of any dispute, difference or question arising between the
parties in connection with this Agreement or any clause or the con-
struction thereof, then and in every such case, unless the parties concur
to the appointment of a single arbitrator, the difference shall be referred
to three (3) arbitrators; one to be appointed by each party, and the third
being nominated by the two so selected by the parties, or if they cannot
agree on a third, by the American Arbitration Association. In the event
that either party within one month of any notification made to it of
a demand for arbitration by the other party, shall not have appointed
its arbitrator, such arbitrator shall be nominated by the American Arbitration Association. The arbitration shall take place in Philadelphia,
Pennsylvania. The arbitrators must base their decision with respect to
the difference before them on the contents of this Agreement and the
attachments thereto, and the decision of any two of the three arbitrators
shall be binding on both parties. The arbitrators shall apply the law of
the Commonwealth of Pennsylvania.

12. PUBLIC STATEMENTS. Any public statements or publicity concerning the existence or contents of this Agreement shall be subject to review
and approval by the other party. TASTEWELL and BIGFOODS will
consult with each other concerning the means by which customers and
potential customers shall be informed of this Agreement.

13. GOVERNMENTAL APPROVALS. BIGFOODS will at its own expense
apply for and obtain the approvals of such governmental and regulatory
entities as necessary to the marketing and sale of the Licensed Goods in
the Territory. TASTEWELL will furnish necessary technical support to
assist in obtaining any such approvals.

14. GENERAL PROVISIONS

14.1 This agreement sets forth the entire agreement and understanding between the parties hereto relating to the subject matter hereof, and supersedes any prior or contemporaneous oral or written representations, inducements or promises not contained herein.

14.2 No amendment or modification of this agreement shall be valid or binding unless the same shall be made in writing and signed on behalf of each party by a duly authorized representative.

14.3 This Agreement and all rights and obligations herein shall be binding upon and inure to the benefit of and be enforceable against the parties and their successors or assigns. TASTEWELL and BIGFOODS shall make no assignment, pledge or hypothecation of this agreement or its performance thereunder without the express written permission of TASTEWELL and BIGFOODS.

14.4 The failure to enforce any of the terms and conditions of this agreement by either of the parties hereto shall not be deemed a waiver of any other right or privilege under this agreement or waiver of the right thereafter to claim damages for any deficiencies resulting from any misrepresentation, breach of warranty, non-fulfillment of any obligation of any other party hereto.

14.5 If any term or provision of this agreement is held to be invalid or unenforceable by reason of any rule of law or a public policy, this agreement shall be deemed amended to delete the term or provision so held to be invalid or unenforceable therefrom and all other remaining terms and provisions of this agreement shall remain in full force and effect. Provided, however, if the invalid or unenforceable provision contains a material term or condition of this Agreement then either party shall have the right to terminate upon five (5) days prior written notice following the determination of such invalidity or unenforceability. If any provision is inapplicable to any circumstance, it shall nevertheless remain applicable to all other circumstances.

15. NOTICE

15.1 Any notice or statement by any party shall be deemed to be sufficiently given when sent by receipted facsimile with a copy by prepaid, trackable overnight delivery, to the notified party at its address set forth above and to its counsel. These addresses shall remain in effect unless another address is substituted by written notice.

[Add Names and Addresses]

In witness whereof, the parties hereto have caused this agreement to be signed, sealed and delivered on the date indicated above.

BigFoods, Inc.

_____   _____

Date   BY:
                                             Name:
                                             Title:

                                             Tastewell Industries

_____   _____

Date   BY:
                                             Name:
                                             Title:

Index